Francis
Keep up the good
work!

Henry

FIELD HYDROLOGY IN TROPICAL COUNTRIES

9 October 2000

DEDICATION
This Introduction is dedicated to the staff of the Physics Division of the former East African Agriculture and Forestry Research Organisation (EAAFRO) based at Muguga and elsewhere in Kenya, Tanzania and Uganda

FIELD HYDROLOGY
IN TROPICAL COUNTRIES

A practical introduction

HENRY GUNSTON

INTERMEDIATE TECHNOLOGY PUBLICATIONS 1998

Intermediate Technology Publications
103/105 Southampton Row, London WC1B 4HH, UK

A CIP record for this book is available from the British Library

© Institute of Hydrology, Crowmarsh Gifford, Wallingford,
OX10 8BB, United Kingdom 1998

ISBN 1 85339 427 0

Tel:

Typeset by J&L Composition Ltd, Filey, North Yorkshire
Printed in the UK by SRP, Exeter

FOREWORD

Among global ecologists there has recently been some consistency in the views that shortages of water supplies in the tropical developing countries will be the first of the crises arising from their rapid population growth. This is due only in part to consumption by the growing numbers of people and livestock. A more serious threat is the increasing damage to the upper watersheds. As uncontrolled subsistence farming spreads uphill into steeper lands, both soil erosion and flood flows increase. Where tropical lands developed under Western influence most of the surviving forests are due to the early creation of protective reservations in stream source areas.

As populations increase in the lower lands of tropical river basins the need for the management of water resources becomes more critical. *The indispensable key to water resource management is measurement.* Henry Gunston's small book on the practicalities of field hydrology is therefore timely. While an increasing part of the literature is given over to computer modelling it is easy to forget that most of the catchment data is from high latitude countries, principally from the USA. Reliance on ecologically irrelevant data can give only erratic results for the tropics.

The author gained his early field experience in the East African series of watershed experiments on the hydrological consequences of land-use changes (Pereira 1989). A survey of world data by UNESCO (Bruijnzeel 1990) quoted these experiments as the main source of reliable quantitative evidence for such effects in the humid tropics. Henry Gunston's subsequent long and wide-ranging experience in tropical field hydrology gives him the authority to present these practical guidelines for the essential measurements on which the management of water resources must depend.

Sir Charles Pereira, FRS

Bruijnzeel, L. A., 1990. *Hydrology of moist tropical forests and the effects of conversion: A state of knowledge review* UNESCO International Hydrology Programme (IHP), Paris.

Pereira, H. C., 1989. *Policy and practice in the management of tropical watersheds*. Westview Press, Boulder, Colorado.

PREFACE

This book is a practical introduction to – not a formal manual or textbook on – tropical field hydrology. It is based on my own experiences as a field hydrologist working in tropical countries during thirty years since I joined the staff of the Institute of Hydrology (IH) at Wallingford in the United Kingdom in March 1968. My work has taken me for significant periods to Ecuador, Kenya and Sri Lanka, and on shorter consultancies and projects to Botswana, India, Niger, Somalia, South Africa, Tanzania, Thailand, Uganda and Zimbabwe. I am currently involved in a field study of the water resources of the remote island of Saint Helena in the South Atlantic Ocean.

I am very grateful to the Engineering Division of the British government Department for International Development (DFID) – and to the department which it succeeded in May 1997, the Overseas Development Administration (ODA) – for support for the production of this text. Initial funding came via an ODA project titled 'A manual for practical field hydrology in tropical countries'. The Department has also, in co-operation with the various organizations in partner countries, funded the majority of the overseas projects on which I have worked.

Special thanks go to those involved in field hydrology in tropical countries (at all levels), who have so willingly shared their expertise with me. Also to many colleagues based at Wallingford with the Institute of Hydrology, HR Wallingford and the British Geological Survey. I would particularly thank Jim Blackie, Terry Marsh, Mark Robinson and Paul Rosier for their comments on the text. The late Professor Norman Hudson of Silsoe College, Cranfield University – whose own practical guides on field engineering and soil conservation have provided many useful ideas – encouraged me to write this Introduction. I am grateful to have used notes and tables from his own *Field engineering for agricultural development* in Appendix 3.

I owe a special debt to Sir Charles Pereira, who has kindly contributed a Foreword, Dr Jim McCulloch and Jim Blackie. Sir Charles set up the experimental watershed studies in Kenya, Tanzania and Uganda on which I started my overseas career in 1968. Jim McCulloch, as Director of the Institute of Hydrology, recruited me and supported my work in East Africa. Jim Blackie (aided by the experienced staff of the Physics Division of the East African Agriculture and Forestry Research Organisation – EAAFRO) took me into the field and introduced me to the pitfalls and pleasures of tropical field hydrology.

On the production side, I have welcomed the support and enthusiasm of Neal Burton and his team at Intermediate Technology Publications. The majority of photographs are my own, but I have been pleased to use a number of views by David Kirby. Finally, thanks to my wife Anne and sons Tom and Rob, who have bounced many a mile in Land Rovers on field trips during my overseas postings.

Henry Gunston, Wallingford, March 1998

CONTENTS

SECTION 3: SOURCES OF FURTHER INFORMATION

Any numerical analysis for
the planning and management of water resources
can only be as good as
the quantity and quality of
the field measurements on which it is based.

Hydrological data collection is a
means to an end
(the better management of water resources)
not *an end in itself*
(simply collecting numbers for the sake of it).

Always try to make hydrological sense of
what the readings are indicating.

SECTION 1
PLANNING AND MANAGEMENT

Why is field hydrology so important?

The study of hydrology provides the foundation for the proper management of water resources in any country. Hydrology itself is founded on a regular supply of good quality measurements of rainfall and of the flows in rivers and streams. Many other measurements are needed for water resources planning – including rates of evaporation of water from land and water surfaces, and well water levels which indicate the depths of water tables below the ground surface.

This book focuses on the measurement and estimation of rainfall, streamflow and evaporation.

Field hydrology concentrates on the following activities:

○ planning instrument networks
○ installing and maintaining equipment
○ co-ordinating the regular collection and recording of measurements
○ safe movement of field measurement records to central offices.

Sites at which hydrological measurements are made are known as field stations. When the records of measurements made at field stations arrive at regional or national hydrological centres, they are checked and safely stored, either as written records or in computer archives – usually both. Once checked and archived, these hydrological measurement records from field stations are ready for use to assist the better management of national, regional or local water resources.

It is essential to establish and maintain field stations at as many sites as possible, and to put every effort into collecting continuous measurement records over long periods. This is especially true in parts of a country where further development of water resources is likely to take place in the future.

When the funding for a long-awaited water resources project finally becomes available, it is too late to discover that no continuous hydrological records are available for the area.

People and teams – not just instruments and measurements

Field hydrology is a *team* activity in which the people who collect measurements and maintain equipment are just as important as the hydrological instruments. Those involved may be:

○ part-time observers with raingauges near their homes
○ members of mobile river-gauging teams
○ technicians servicing and repairing instruments
○ senior managers of national hydrology programmes.

Keeping up the enthusiasm and commitment of observers and field teams, and maintaining the standards of records stored in national data archives, are continuing challenges to the managers of hydrological programmes.

An introduction – not a manual or textbook

It has already been noted that this is a practical introduction to – not a manual or a textbook on – tropical field hydrology. The book is aimed at those who have already studied some of the principles of hydrology, but who have not yet gained wide experience of practical, day-to-day field operations. It has been written with the following points in mind:

Not too long
I have deliberately attempted to limit the overall length of text. The World Meteorological Organization's *Guide to Hydrological Practices* (discussed in more detail below) has 735 pages in its latest edition, but that is a book to be referred to, not read from end to end.

A mix of operating principles and practical tips

I have combined descriptions of the operating principles of the various instruments and techniques used in field hydrology with practical tips drawn from my own experiences when working in tropical countries. Some of these tips may only seem to be important when faced with the immediate circumstances to which they refer. I have included them, however, to help create the 'feel' of handling day-to-day practical hydrological problems in the field.

Operating rather than installing instruments

Newcomers to field hydrology start by helping to operate existing instrument networks. Taking charge of installing equipment comes later. The focus here is therefore on instrument operation in the field rather than the selection of sites and the installation of new equipment.

'Appropriate Field Hydrology'

National hydrological services all over the world are always seeking the most cost-effective ways of collecting data and extending instrument networks. In Chapter 2 the idea of 'Appropriate Field Hydrology' is presented. This is based on the more widely known *Appropriate Technology* approach, which aims to match the needs of a job with the complexity and cost of the methods used to carry it out.

Focus on national water resource management rather than irrigation

The most precise measurement and management of water in tropical countries is often linked to irrigation. However, there is already a very extensive range of publications available on the hydrological aspects of the water balances of irrigation schemes. The main focus of this book is on the operation of networks of hydrological field stations on a country-wide basis, to assist national and regional water resource planning. Although some aspects of field hydrology which are closely related to irrigation water management are mentioned, for detailed coverage the reader is referred to the specialist irrigation literature.

Guidance on further reading

Hydrological publications which I have myself found useful are listed in Appendix 1. In addition, valuable guidance on field hydrology techniques is available internationally through the World Meteorological Organization (WMO) and the International Organization for Standardization (ISO), of which most countries of the world are Members. Further details appear in Chapter 1 and in Appendix 2.

Measurement units and equations

Following the 'introduction' approach to field hydrology, full numerical information has not been included here up to the level of a manual or textbook. However, appropriate notes on measurement units, together with some basic hydrological equations, are presented. Shorter entries appear in text chapters and there is an overall collection of numerical information in Appendix 3.

1
INTRODUCTION

Tropical field hydrology

This book aims to give guidance to those work-ing in tropical countries, where special features of hydrology include:

○ streams and rivers in drier areas which do not flow all the year round
○ intense rainfall rates during storms, which can generate sudden flood flows in rivers
○ measurements taken mainly at manually-read stations operated by observers
○ few permanent streamflow gauging struc-tures, such as weirs or flumes
○ an absence of severe cold weather condi-tions (such as ice formation on rivers and heavy snowfall) except in the highest mountainous regions.

General guidance on working in the field

Before discussing the instruments and methods used in tropical field hydrology, the overall operation of data* collection programmes must be considered. Field work in remote areas pre-sents challenges on staffing, communication, transport and safety. Chapter 3 is therefore focused on the broader scene of working in the field. The general principles of inspecting and maintaining equipment in good working order are an essential part of that 'broader scene' of field hydrology, where long-term continuous records are of great importance.

Good links between headquarters and the field teams

Problems can arise in hydrology if commun-ications are poor between the staff who are involved in field operations and those who work full time at headquarters offices. An 'us and them' culture can develop:

Field station observer: 'I only received these recorder charts after I telephoned headquar-ters *five times*. ...'

Headquarters data archivist: 'That observer forgot to wind up the raingauge recorder clock *again*. ...'

To prevent harmful divisions building up between headquarters and field teams, staff who check and archive data should be encour-aged to visit the field stations from which the data have come. Headquarters staff can inspect sites and equipment and talk directly to observers. In return, field station observers should visit 'head office' to see how the data they have collected are used. Observers often live and work in remote areas, and it will give them extra pride in their day-to-day activities if they are shown how important their routine data collection work is to national water resource planning.

Make use of (human) memories

Despite all the best efforts made when storing hydrological data in archives, either as paper records or in computer-compatible form, some very useful information may remain only in the memories of long-serving staff. They may be

---

* *Data* is the word used to indicate a collection of numerical measurements. In correct use it is a plural word (singular: datum), because there is always more than one measurement in a set of data. 'Data *are* stored' is therefore right, but 'data *is* stored' is wrong.

field station observers or headquarters data archivists, but they should always be encouraged to pass on – and especially to write down – details of past station performance, of extreme hydrological events and of any alterations made to instruments or recording procedure; once they retire it will be too late.

'Odd' records can be true records of 'odd' events

Headquarters hydrologists who analyse data as part of water resource development projects can help greatly with maintaining and improving the quality of archived data. Analysis often highlights what appear to be 'odd' data values in the records. When followed up at the field station, it may be discovered that these values resulted from instrument or recording problems. However, unless clear explanations can be found, these apparently 'odd' values must not be removed from, or altered in, archived records. They may be true values of important 'odd' hydrological events.

Hydrology and meteorology – co-operation is essential

In many countries the responsibility for collecting and storing data on rainfall, streamflow and evaporation is shared between national organizations concerned with hydrology and those concerned with meteorology. Both hydrologists and meteorologists have an interest in collecting rainfall data, but measurements of flows in streams and rivers are normally collected by national hydrological services. The estimates of evaporation from land and water that are needed for water resources planning are normally calculated from climate measurements made at meteorological (met.) stations, which are discussed in detail in Chapter 6. Simple met. stations, equipped with raingauges and an evaporation pan, may be operated by the hydrological service. The more complex stations, where a range of climatic measurements are made, are often operated by national meteorological services. Field hydrology pro-

grammes therefore rely on good co-operation between the national hydrological and meteorological services. Important areas of joint activity include:

○ planning raingauge and meteorological station networks
○ sharing collected rainfall (and other climate) data
○ co-operative programmes for field station inspections.

'Met.' for 'meteorological'

To save further repetition of the word 'meteorological', from now on it will be generally replaced by the shorthand form 'met'. In addition, to avoid repeating the phrase 'hydrological and meteorological', the terms 'field hydrology' and 'hydrological field station' will be used to include activities covering measurements in both specialist areas.

Electronic data collection and transmission

Most hydrological measurements taken in tropical countries are written by hand on record forms or in notebooks, or indicated by pen traces on recorder charts. However, electronic data collection is being introduced progressively. The measuring instrument at the field station (e.g. raingauge or river water level sensor) produces a data record in an electronic form which is stored within a separate unit called a data logger which is placed nearby. The earlier designs of data logger stored data on magnetic tape cassettes, like those used to record and play music. However, the mechanisms which moved the tape forward after each reading did not always operate well under extreme climatic conditions of heat, cold or wetness, so 'solid state' data memory modules (with no mechanical moving parts) are now usually used in data loggers.

Collecting electronic data

The stored data can be collected at the recording station during the visit of a field team in the following ways:

○ by removing a detachable solid state electronic memory unit from within the data logger and replacing it with an 'empty' memory unit in which new data can be stored

○ by down-loading the data stored in the data logger directly into the data memory of a portable computer (PC), using a temporary connecting cable which links the PC to the data logger.

In addition, hydrological information can now be transmitted from field sites to central offices through high technology telemetry links – including telephone connections (by cable or radio) – and by the use of satellites orbiting far above the earth's surface.

Advantages of using electronic recording instruments include:

○ Telemetry links (telephone line, radio or satellite) make it easy to collect data, with site visits required only for servicing and maintenance purposes.

○ Instruments can be operated at remote sites where it is difficult to station an observer to read a manual recording station – it is becoming more difficult to find staff who will work as field station observers for relatively low wages in remote areas.

○ These instruments can make measurements at very frequent intervals which, when integrated over 24-hour periods, produce much more accurate data than can be obtained from manual readings by observers, which may be taken only once a day.

○ The costs of some 'manual' station instruments (such as sunshine recorders and 'clock and drum' recording raingauges) tend to be rising, whilst the cost of electronic circuit 'chips' – and thus of electronic recording instruments – tends to be falling.

However, Appropriate Field Hydrology thinking, discussed in Chapter 2, must be applied to the selection, installation and operation of electronic recording instruments. The benefits and costs of using them must be balanced against the benefits and costs of using simpler 'manual' instruments.

Where data are recorded (and possibly transmitted) automatically, field sites may be unstaffed except when maintenance of the instruments is carried out. The installation and maintenance of instruments which record or transmit data electronically is normally the responsibility of specialist engineers and technicians. However, field hydrologists are often expected to carry out routine operations, such as changing electronic memory modules or down-loading data from loggers into portable computers. They will also make visual checks that the instrument appears to be in good order, and that any mechanical parts are operating smoothly.

Good field hydrology practices are just as essential when using electronic data collection and transmission as when using older style 'manual' methods. At unstaffed sites instruments must be installed and maintained to high standards as there is no observer present to check equipment daily.

Field hydrology guidance from WMO and ISO

It has already been mentioned that two international organizations provide useful guidance on field hydrology. The World Meteorological Organization has produced a particular useful *Guide to Hydrological Practices*, the latest (1994) edition of which runs to 735 pages. This Guide not only covers in greater detail most of the subjects discussed in this book, but also deals with a wide range of other topics, including measurement of snow cover, sediment discharge, soil moisture, groundwater and water quality.

Another major source of field hydrological advice is the WMO *Hydrological Operational Multipurpose Sub-programme (HOMS)*, which is cross-linked with recommendations given in the Guide mentioned above. Experts from many countries contribute components of hydrological information, such as manuals, training videos or computer programs, to

HOMS, which has a central office in the WMO headquarters in Geneva, Switzerland. Once checked as being of the high standards demanded by WMO, details of these components are circulated to HOMS National Reference Centres, which exist in most countries of the world. Details of the *WMO Guide to Hydrological Practices*, and of how to obtain copies of HOMS components, appear in Appendix 2.

The International Organization for Standardization (ISO), publishes detailed International Standards for carrying out particular activities, including streamflow measurement. These describe standard methods of working and construction, and they are prepared and approved by experts from many countries, so that the same job can be carried out in the same way worldwide. Streamflow measurement techniques are included among the ISO International Standards on liquid flow measurement in open channels. Details of these Standards, and how copies of them can be obtained, are given in Appendix 2.

Catchment, watershed, divide and basin

Finally, the words *catchment, watershed* and *basin*. In British hydrology practice and textbooks, the area of land that drains into a river upstream of a certain point is a catchment, and the boundary which separates one catchment from others around it is a watershed. To hydrologists trained in the USA, however, a watershed is the area of land draining (which the British call a catchment) and the line of separation of watersheds is a divide. The word basin is generally used to indicate the total catchment (or watershed) of very large river such as the Mekong or the Zambezi.

2
'APPROPRIATE FIELD HYDROLOGY'

In most countries, national hydrological data collection organizations never have as much funding as they would like to operate, maintain and extend their networks of instruments. An approach is offered here to help decide how best to use the funds that are available.

In skilled work there is a saying: choose the right tool for the job.

Doing any job properly, in field hydrology as in the rest of life, requires choosing the right tools and materials to complete the work to the standard that is required, and using time, energy and resources as efficiently as possible. The description 'Appropriate Technology' has been used to indicate a realistic, practical, questioning approach to selecting 'the right tool for the job' and to matching the cost and complexity of inputs to the precise needs of the job. For example, is it essential to use an expensive, imported piece of equipment to do this job, or will a cheaper, home-produced product do the job just as well? The answer may well be that the imported equipment is required, but the important point is to ask the question before making a final decision.

The thinking behind appropriate technology, and the closely related 'intermediate technology' approach, can be found in the following two books (for details see Appendix 1):

○ *Small is beautiful* by Dr E F Schumacher
○ *Appropriate Technology – Technology with a human face* by Professor P D Dunn.

Building on this questioning and resource matching approach, the idea of 'appropriate field hydrology' can be created.

A practical example: the selection of recording raingauges

As an example of the use of the Appropriate Field Hydrology approach, the selection of recording raingauges to produce continuous records of rainfall will be discussed.

There are two basic types of recording raingauge, shown below and overleaf.

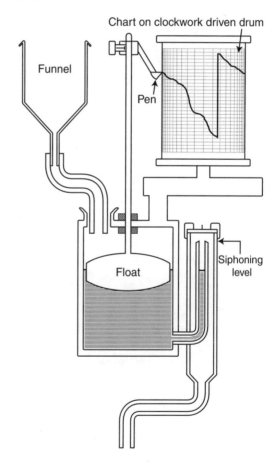

Fig. 1(i) Recording rainguage: Clock and chart siphon type (not to scale)

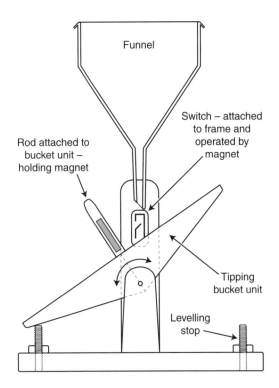

Funnel

Switch – attached to frame and operated by magnet

Rod attached to bucket unit – holding magnet

Tipping bucket unit

Levelling stop

Fig. 1(ii) Recording rainguage: Tipping bucket electronic type

Clock and chart gauges

These older designs rely entirely on mechanical operation and are comparatively simple to maintain. However, the charts on which data are recorded have set time periods, at the end of which the gauge *must* be visited. Transmission of data through telemetry links is not possible from this type of gauge. These designs operate as follows:

a) rain from the collector funnel runs down into a chamber with a float in it
b) the float is connected by a vertical rod to a pen
c) the pen writes on a chart which is wrapped round a drum
d) the drum is rotated by a clock, thus moving the chart
e) when rain falls, the float rises and the pen records a rising trace on the chart
f) when the chamber is full, it is automatically emptied by a siphon system and the float falls to the bottom of the chamber

g) linked to the float, the pen falls quickly to the bottom of the chart, ready to rise again with the float as more rain falls.

The following points are important:

o at the end of the time period of the chart (often seven days) an observer must visit the gauge to change the chart
o the observer's visit also allows the operation of the raingauge to be checked
o the pen trace on the chart has to be 'translated' to provide data in numerical form for archiving and analysis
o the 'translation' may be done by hand, or using an electronic chart-reader, which uses light signals to detect the shape of the pen trace.

Electronic recording tipping bucket gauges

Newer designs of recording raingauge convert rainfall data into electrical pulses, one of which is produced each time a predetermined depth of rain falls into the gauge. These data can be stored electronically in solid state memory modules in data loggers, as described earlier. Data may also be transmitted through telemetry systems, which allow information to be sent immediately over long distances. This is particularly valuable for flood warnings, when upstream rainfall and river level measurements can be sent the moment they are recorded directly by telemetry to flood defence offices.

The basic principles of operation of tipping bucket gauges are as follows:

a) mounted below the collector funnel is a tipping bucket unit, which operates like a see-saw in a children's playground, working around a horizontal centre pivot
b) on either side of the centre pivot is a small container (bucket), precisely shaped so that it will tip the see-saw when an exact weight of rainwater has entered from the collector funnel
c) if the tipping bucket unit is calibrated to operate every time 0.5 mm rain falls, the weight of rainwater required to fill a bucket and tip the see-saw will be the weight of the

Mechanism of a 'clock and chart' recording raingauge, with upper casing and funnel removed. Note pen rod rising from the float chamber and the top of the siphon tube in front of it (Sri Lanka).

volume of the cross-section area of the rain-gauge rim multiplied by 0.5 mm

d) each time a bucket empties, the water drains away – although it can be collected in a storage container below the gauge for checking purposes

e) attached to the see-saw unit directly above the centre pivot is an upright rod which carries a magnet – this rod moves from side to side as the see-saw tips

f) fixed to the frame of the gauge above the centre pivot as another upright rod which carries an enclosed switch, with its contacts sealed within a glass envelope

g) each time the see-saw tips, the magnet moves past the fixed switch mounting, closing the switch for a short period

h) an electronic recording circuit counts each closure of the switch

i) each switch closure records a bucket tip representing the calibration rainfall value – typically 0.5 mm.

Periods of data recording by tipping bucket recording raingauges can be much longer than a week, depending on the capacity of the electronic memory contained in the data logger to store the signals generated by bucket tips.

Applying the appropriate field hydrology approach

Imagine that one of the older type clock and chart gauges, sited at a station where there is an observer, needs major repair. The appropriate field hydrology approach requires that questions are asked:

o Should the older type raingauge be repaired locally, or should a new electronic recording gauge be imported from overseas?

o How much will it cost to purchase and import the electronic recording raingauge?

o Is it becoming difficult to obtain spare parts (such as the clock which moves the chart) for the mechanical gauge?

o Will spare parts need to be imported to maintain the electronic gauge?

o Which raingauge is likely to produce the most reliable, continuous records?

o Which will give records which are simplest to check and store in a data archive?

o If the electronic raingauge is installed, will there still need to be an observer on site?

These questions, and others, will result in further thinking:

o Rainfall data which come from the raingauge via a solid state module in a data logger can be fed directly into computers.

o Chart traces from clockwork-driven recorders need to be translated into numerical form before they are fed into the computer – either by hand or using an electronic chart-reader.

o Clocks, pens and ink can give problems – clocks stopping, pens drying or ink spilling.

o If a chart recorder is not working correctly, a site inspection normally shows that there is a problem.

o If an electronic recording system has failed to operate, there may be no indication on site that it has done so.

o The failure of electronic data collection may become clear only when the storage module from the data logger is decoded at head office.

o If electronic recording tipping bucket raingauges are introduced, arrangements must be made to maintain and repair them and their associated data loggers.

This example of decision-making on recording raingauges has been presented to show that the appropriate field hydrology questioning approach is a way of thinking which helps:

o to decide which equipment is best matched technically to carry out any particular job

o to plan and operate field networks and staff requirements as effectively as possible when funding budgets are limited.

3

WORKING IN THE FIELD

An overview of field work

This chapter covers the operation of field data collection programmes in general terms. Although measurement practices covering rainfall, streamflow and evaporation will be mentioned as examples of field activities, each of those areas will be discussed in more detail in the chapters that follow. The focus here is on three important aspects of working in the field:

○ field programme operation, including working with field teams and observers
○ the operation of vehicles, including what equipment to carry in them
○ fieldwork safety.

Although it is sometimes necessary for field teams to camp out in tents at remote field stations, the equipment and operational needs of camping itself are not dealt with in this chapter.

Part 1: Field programme operation

Introduction

In addition to travelling in vehicles to and from remote places, the management of field hydrology programmes presents many challenges, including:

○ organizing and motivating field teams to work in remote, inhospitable and sometimes dangerous areas
○ supporting and motivating observers based in those remote areas, many of whom may not be direct employees of the data collection organization
○ training and supervising all field staff so that instruments are read correctly at pre-

cise times each day and the data are correctly recorded
○ encouraging observers and field teams to continue working (within safety limits) during extreme rain and flood events, when they may be asked to take additional readings under difficult conditions at any hour of day and night
○ organizing instrument maintenance and the supply to observers of renewables such as charts, data recording forms, field notebooks and sunshine recorder cards.

Some field stations are operated by an observer who lives nearby. Others have no observer and the instruments operate unattended during periods between visits by field teams. Some observers will be full-time employees of national hydrological or meteorological departments. Others may be farmers or schoolteachers who have volunteered to read a raingauge once a day, and who may or may not receive formal payment for doing so. Data collection programmes for a group of field stations are supervised by hydrological field teams. The members of these teams are usually based at the headquarters, or at a regional office, of the national hydrological organization.

Regular inspection visits to field stations

It is essential that field stations are visited regularly so that records can be collected, equipment checked and operations discussed at first hand with the observer. There is a danger in putting off field inspection visits, especially when station record sheets arriving by post at headquarters look satisfactory – or when there

A woman observer being instructed in operating a long-term storage raingauge (Ecuador).

14

is a shortage of vehicles available for field teams to use. However, it is most unwise to allow a timetable of regular field station inspection visits to be changed.

The whole purpose of field hydrology is to produce continuous, unbroken records of good quality data. If you can identify and solve a possible instrument problem during a routine field inspection *before* the instrument fails to operate correctly, you have helped to maintain the unbroken record of data from that site.

Will the observer be there when you visit?

Except at important meteorological stations, such as those at airports, even observers who are full-time employees are likely to have duties away from the field station during the working day. Part-time observers who only read a raingauge each morning obviously have other commitments. It is therefore most important to confirm with an observer that he or she will be on site at the time when a field team wishes to visit the station. It can be very frustrating to arrive at a station towards the end of a morning, only to be told that the observer will not be back until the evening. However, if there are no station reading responsibilities in the afternoon, and if the inspection visit was not arranged in advance, you cannot blame the observer for not being on site. Either agree a definite date and time for a next visit each time you are at the station, or use telephone or letter post to advise when the field team plans to visit.

Never rush a site visit

Always take time to talk with the observer, and listen for any comments on unusual readings or hydrological events. A field team leader must always check through any completed data sheets and other records which are ready to be sent to headquarters, and discuss them with the observer before leaving the field station.

The more distant a measurement value becomes, in time or place, from the moment and the site at which it was taken, the more risk there is of any errors in it passing unnoticed.

Take time to visit all the instruments on site and to discuss with the observer how well they are operating. By visiting the instrument the observer may be reminded of some problem which has not been noted down in the station record book.

Station record books – essential support to numerical records

Many of the instrument readings taken by an observer at a field station will be written directly on to standard data collection forms or field books. Examples are daily morning readings of rainfall, river water level and air temperature. However, a great deal of information on hydrology and instrument operation at field stations does not come in a form that fits neatly on to the standard data recording sheets. Examples are:

o When a flood occurs, lines of material carried by the flood waters are often left along river banks, indicating the highest level that the flood water level reached. The observer should record their position after the flood.

o The free flow of wind across a meteorological station (and therefore the correct value of evaporation estimates) is affected by any trees growing nearby. On which day were the trees cut back?

o Older designs of rainfall and water level recorders contain clocks which move the drums that carry the charts. When was each recorder clock last serviced or replaced?

o On which day did the elephant break down the meteorological station fence and drink from the evaporation pan?

Make sure that observers are provided with station record books in which to record this vital extra information. Discuss with them the sort of extra details which need to be noted, and tell them how important it is to write down the information as soon as possible, before it is forgotten. Observers should be encouraged to use these record books as a station diary, although

Daily and recording raingauges in fenced enclosure, with enthusiastic observer (Sri Lanka).

they need record information only on days when there is anything significant to record. Do not remove the record book from the station unless it is completed, and properly replaced by a new book.

Check everyone's handwriting – including your own!

Emergency! There has just been a peak flood on a major river and it is essential to know the value of the highest water level. An observer was on site and was able to read the river level gauge – but no-one can read the entry that has been written on the field data form....

Make sure that all field hydrology information is written clearly on record forms and in field notebooks, using block letters if necessary. Look out for a 2 that looks like a 7, and a 3 that looks like a 5, and vice versa.

Remember that the staff who enter field data into computers at headquarters are usually not hydrologists. When punching large quantities of numbers into computers, the process becomes almost automatic. This means that if a number is mis-read from a field form because it has not been clearly written, the false value may go unnoticed into the data archive.

If an observer does not write clearly (especially numbers), discuss the matter and make a point of commenting favourably later if it improves – or unfavourably if it does not!

Can others read *your* handwriting?

Timekeeping for observers

A correct record of the time when an instrument is read is often as important as the numerical value of the measurement itself. An observer must therefore have a reliable watch or clock on site to ensure correct timekeeping. If the department is not able to provide a watch or clock for the observer officially, it may help to arrange a personal loan so that a watch can be bought and the cost paid back in instalments.

Whatever timekeepers observers use, check that the time they are keeping is correct. Time

checks provided by radio and television stations are very useful for checking clocks and watches each day.

Checking stocks of renewable items at field stations

A wide range of renewable items is required to keep data collection progressing smoothly at field stations. Requirements linked directly to measurements of rainfall or streamflow will be covered in detail in the following chapters, but examples include:

o   data recording forms and notebooks
o   charts and ink for recorders
o   sunshine recorder cards
o   batteries and data logger memory modules for electronic reading instruments
o   pens, pencils, writing paper, diaries and wall calenders
o   expense claim forms for observers.

The full list of renewables needed for an extended trip to visit a number of field stations will contain many items.

Where stations are staffed by observers, a regular stock of renewable items must be maintained on site, and a minimum number of each item to be kept at each station must be set. Field teams must carry stocks of renewable items and during every field station visit the team leader and the observer must together check the stock of each item. Minimum numbers can then be made up from the stock carried by the field team.

Field notebooks for team leaders

In the same way that permanent record books are needed at field stations, field team leaders must always carry a field notebook to record additional information, for example:

o   The river level gauge board has been damaged by a large tree trunk carried down during a flood and must be replaced – check the station record book for the date when the damage occurred.

o How many water level recorder charts and sunshine recorder cards will need to be delivered on the next field team visit to maintain the minimum stock at the station?

o Over what period does the observer wish to be on leave and what alternative staffing arrangements must be made?

The field notebook can also be used to transfer important information from the station record book to the station record files at headquarters.

What if you lose your field notebook? Out on site there is a risk that your field notebook (and all the vital data in it) will become lost, dropped into a river or left behind in some remote place. Always have this risk in mind, and ensure that fresh information is 'downloaded' from the notebook as soon as you return to headquarters, as described in the next section.

Routing the notes from field trips at headquarters

The vital information collected in field notebooks is of no lasting value unless it is routed by field teams to the right destinations back at headquarters. There are three main routes to follow:

o Information related directly to quantity and quality of data being collected goes to the station record file which will be kept at headquarters for each field station.

o Information on repairs, work to be done and supplies needed at the station goes on to a Field Team Action List so that preparations can be made before the next field team visits the station.

o Information on personal matters raised by observers (some of which may be confidential) will be discussed with their line managers and the personnel department, as appropriate.

As mentioned in the last section, information from field notebooks should be transferred as soon as possible. Not only is it more efficient to sort out matters from a field trip as soon as pos-

sible, but if a field notebook is lost on the next trip, valuable information will not be lost with it.

Keeping the observer's goodwill

An observer is required to read a river level, a raingauge or a meteorological site in a proper manner at regular hours. To do this, he or she not only needs good equipment, but also the motivation to do the job properly. Working in remote places can produce a sense of isolation. To an observer, headquarters seems far distant. It is probably in the country's capital city, where everyone seems far too busy to take any interest in his or her work or life.

The leaders and other members of field teams therefore have a special responsibility to win, and to keep, the observer's goodwill. Remember that the quality of the hydrological records taken at the station depends on the observer's motivation, unsupervised, to collect them properly. His or her motivation, in turn, is dependent on the amount of enthusiastic encouragement received from members of visiting field teams.

'My salary has not been paid'

Observers, like all of us, become upset if salaries and allowances have not been paid promptly. For staff based at headquarters, access to accounts and personnel departments is comparatively easy when personal salary and allowance problems need sorting out. Distant observers, however, may need visiting field team leaders to argue for them at headquarters if they have not been paid. For payments due other than salary, if the sums are not too large, it can help if the visit team leader makes the payment in cash direct to the observer (and obtains a signed receipt) during a field station visit.

Pay serious attention to any pay problems of observers. Isolation can magnify grievances against headquarters and produce disillusionment which shows up through poor quality work. Where the observer is officially due other supplies, such as foodstuffs or wellington boots, as part of terms of employment, make sure those supplies are correctly delivered as well.

Personal loans

If observers are in a difficult financial position, possibly due to delayed payments from head-quarters, they may ask field team leaders for personal loans. As hydrological departments usually have no formal method of making loans, these are an individual matter between the visiting team leader and the observer. It is essential that the terms of the loan are written down, with both those involved having copies. Personal judgement, tested over time, will establish those who work hard to repay loans, and those who do not!

'How are your crops and cattle?'

The discussion so far has mainly focused on the circumstances of observers who are employed full-time by the hydrological department. How-ever, there are also many part-time observers who may be paid small sums (or sometimes nothing at all) to make one or more daily read-ings of rainfall or river level. The hydrological value to the department of the measurements these people take is very high, and every effort should be made to encourage and support them.

Always remember that although hydrology may be first priority in your working life, to a part-time observer who is a farmer or a school-teacher, taking the readings is only one of many priorities in a busy working day. Once again, do not rush a site visit. Take a little time to dis-cuss matters such as:

○ How are the crops growing this season?
○ How are things at the school this term?
○ Is the allocation of irrigation water satis-factory?
○ Are the school pupils interested in the rain-fall readings?
○ How are the cattle doing?

As with full-time observers, make sure that any payments due to the observer are settled promptly. Once again, direct cash payment by the field team leader against a receipt from the observer is often the simplest answer. If no formal payment can be made, think of other ways in which the field team can help. Can you transport something which is needed from the town on your next visit? Perhaps a sack of food too heavy to be easily transported by bicycle.

New equipment – and lost screws

Two final points on working with instruments in the field:

New instruments

If you are involved with introducing a new piece of field hydrology equipment into your country, make sure that it is fully tested before being placed in regular network use. If it is planned to replace existing network instru-ments with those of the new design, keep both types in operation at field stations operating for at least a six-month trial period – or through the next monsoon season. Remember the appropri-ate field hydrology approach, and give up using the old instruments completely only when you are fully satisfied that the new ones will pro-vide reliable, continuous data sets.

Don't lose the screws

A danger when servicing instruments in the field is that screws, nuts and other small parts will fall to the ground and get lost among stones and grass. Carry a flat tray with raised sides on which to service instruments, and some flat bottomed tins (plastic pots can be knocked over) into which you can put screws and other components as you remove them from the instrument. Do the job on a flat raised surface – the back floor of a four-wheeled drive vehicle can be ideal for this type of work.

Part 2: Operation of vehicles

A basic rule

There is a basic rule of field work: to control the job, control the transport.

Well organized transport arrangements are essential, not only to get the work done but also to prevent field teams being stranded in dangerous circumstances with broken-down vehicles.

The African highway code

A very good, well-illustrated introduction to driving in tropical countries has been produced by United Nations Economic Commission for Africa and the U.K. Transport and Road Research Laboratory. Entitled *The African Highway Code: A guide for drivers of heavy goods vehicles*, two versions are available – one for countries where vehicles drive on the left and the other where driving is on the right. Although some sections of the text focus especially on driving heavy goods vehicles, most of the guidance given applies to all classes of drivers and vehicles. Details of the Code appear in Appendix 1.

Team leaders must control all vehicle movements

Drivers are key members of field teams, and team leaders must have good relations with them. However, it must be clearly understood at the start of any field trip that although the driver may have charge of operating (and probably maintaining) the vehicle, where and when it moves is entirely under the team leader's control.

If a driver has to be sent with the vehicle away from where a field team is working, the team leader must give clear instructions as to the driver's duties, and agree a planned time of return. At remote river gauging station sites, team safety can be at risk if the driver does not return at the expected time, due to having misunderstood the instructions given. Field trip vehicles carry all the team's equipment (including the first aid kit), so if the team becomes stranded, problems can arise:

○ Lost time upsets field trip planning.
○ Not all the maintenance tools and equipment needed on site may have been unloaded on arrival.

○ Food, drink and first aid equipment have probably remained in the vehicle.

Don't lose your driver!

If a field station visit is planned to take a long time, the driver may wish to take a break to get food and drink, to do some shopping if there is a town or village nearby, or to visit a relation. The same operating rules apply as when a team leader sends a driver on official business. The team leader and driver must clearly agree a time when the driver has to be back at the station, with the vehicle ready to move on.

Keeping the vehicle rolling

The vehicle is the team's lifeline while on a field trip, so team leaders must check out its operating state with the driver before the trip starts. Do not be afraid to delay the start of a trip, especially when travelling to remote areas, if you are not fully satisfied that the vehicle is in proper working order. Breakdowns in remote areas can be annoying and possibly dangerous, so it is very helpful if at least one member of the field team, in addition to the driver, has a working knowledge of vehicle maintenance. Under field conditions a full repair may not be possible, but it will help greatly if the team has the skills to keep the vehicle moving so that it can be driven to the nearest vehicle mechanic's workshop.

Emergency drivers

Although many organizations in tropical countries insist that official vehicles should always be driven by official drivers, it is very useful if another member of the team can drive in emergency – the official driver may fall ill during a field trip.

Keep friends with the transport manager

In most hydrological organizations the responsibility for keeping vehicles in good running order lies with the transport manager, with whom the wise field team leader maintains good personal relations. Always remember that there are 'good' vehicles and 'not-so-good' ones in any transport pool – you wish your field teams to be allocated the good ones!

What to carry in the vehicle

Before starting on a field trip, it is essential to check the equipment carried in the vehicle, which will include:

- equipment needed to carry out the field hydrology work
- equipment needed to keep the vehicle running
- first aid and safety equipment
- drink, food and the team's personal belongings.

Equipment for field hydrology

This can be divided into:

- renewable items to maintain the data collection programme
- tools and equipment for servicing instruments
- equipment and materials for any extra repair and installation work planned during a particular field trip.

Renewable items

The supply of renewable items required at field stations for data collection, such as data recording forms and sunshine recorder cards, has been discussed earlier. In the section on routing field notes at headquarters, the need for a field team action list was mentioned. This list should include details of the likely requirements of the various renewable items which the field team will need to carry on each trip to make up minimum station stocks.

Tools and servicing equipment

The tool kit required for each field trip will be built up from experience with the particular instruments installed. It will include:

- a general tool kit
- tools and spare parts needed for routine servicing of each type of instrument
- additional tools needed for any extra repair

and installation work planned during a particular field trip.

The general tool kit will include various sizes of screwdrivers and pliers, wire strippers, a medium sized hammer, spanners (fixed sizes and adjustable) and plastic electrical insulating tape. Specialist tools will include those required to service individual instruments, including hexagonal Allen keys of various sizes and any non-standard tools.

Requirements for extra work

The field team action list provides a reminder of any special requirements for extra work at a particular field station during the next field team visit. In addition to the renewable items already discussed, there is a need for a field team to carry a range of replacement items for station equipment which may have become damaged. Examples are:

- clocks, pens and clock keys for chart recorders
- thermometers for temperature screens – carefully packed for field travel
- bottles and watertight inner cans for daily raingauges.

Equipment to keep the vehicle running

Field trips are often made to remote sites, far from vehicle mechanics. Field team leaders must therefore check with drivers not only that the vehicle is in full working order, but also what equipment, tools and spares are carried in the vehicle. Once again, ask questions:

- Is the spare tyre fully inflated, without any punctures, and is it worth taking along a second spare wheel?
- Are the jack and wheel nut spanner in full working order?
- What does the vehicle tool kit contain?
- What spare parts are being carried?
- It is normal to carry one can of water – is there a need to carry extra cans?
- Is there a need to carry extra fuel?

Remember to check all cans, whether carrying water or fuel, for leaks before starting. Be very careful when taking petrol in cans in a vehicle. It is unwise to carry cans full of petrol over the front bumper of a four-wheel drive vehicle, as they could burst during a collision, presenting a very serious fire risk.

Detailed answers to the questions listed above will not be given here, but it is far better to obtain answers to them before leaving the headquarters transport yard, rather than end up sitting in a broken-down vehicle which has suddenly attracted the attention of a large, aggressive elephant!

The need for extra tools and spares will depend on the length of the field trip and the remoteness of the stations to be visited. It is always useful to carry the belts which drive fans, water pumps and alternators, as well as spare radiator hoses. With petrol-engined vehicles do not forget to carry spark plugs and contact-breaker sets. This is only an initial list of spares – particular types of vehicle need particular spares. Discuss these matters with drivers and transport managers.

Punctures

Always carry tyre inner tubes and levers to remove the outer tyre casing. For trips across rough country or in any remote areas, a second spare wheel and tyre also should be carried. My own experience is that fitting a spare wheel or a fresh inner tube provides a far better cure for punctures in the field than trying to use puncture repair kits. That does not mean, however, that inner tubes with only small puncture holes (as opposed to tears) cannot be repaired in garage workshops and used again. Always repair a punctured tyre as soon as possible.

Call at the first village or town garage workshop you reach on the field trip after the puncture. Do not wait until the vehicle returns to headquarters. Remember that the most likely time to get a second puncture always seems to be when you are travelling (even for a brief period) without a serviceable spare tyre!

First aid, water and food

First aid skills and equipment are essential for field teams, and they will be discussed in the next section. Drink, food and the team's personal belongings have also been mentioned. When the weather is very hot, especially if travelling to remote areas, it is essential to carry plenty to drink. It is safer to carry drinking water in separate containers from those used to carry water for vehicle and washing use. Large clean plastic bottles are best for carrying water. Plenty of drinking water must be carried, not only to cope with heat during the normal field programme, but as a safety reserve if the vehicle fails in a remote area and some team members have to stay with it. Field team members are normally expert at organizing their food, combining supplies brought from home with food bought at markets and shops along the route. If timing is tight, however, team leaders need to be sure that shopping expeditions do not take too long. (A visit to a bar for a 'quick drink' in the middle of a hot day can seriously delay the field trip timetable!)

Part 3: Safety during field work

Introduction

Field hydrology can be dangerous. Safety precautions are especially important because accidents in remote places are much more difficult to deal with than those that occur in towns, where doctors and hospitals are near at hand. Field team leaders and other team members must be trained in first aid, and a properly stocked first aid kit must be carried in the field team vehicle.

All team members, and especially team leaders, must be always on the alert for possible dangers during field trips. The location of the nearest dispensaries, hospitals and doctors' surgeries to the route of any field trip must be known to both the team leader and driver, especially if the route is one that field teams use regularly.

Road travel always carries risks. Accidents can happen as easily on rough tracks, with no other vehicle around, as they can on busy trunk roads. The need for field team leaders to supervise drivers has already been mentioned, and over-speeding or any other reckless driving must be dealt with severely. Taking vehicles across rivers in flood can be particularly dangerous. If a team member cannot safely wade across the river ahead of the vehicle to check the depth of the water, it is not safe to attempt to take the vehicle across. Further guidance on safety aspects of road travel, including first aid, is given in *The African Highway Code*, which was mentioned earlier.

A more general handbook, *Site Safety in the Water Industry*, has been produced in the UK. Details appear in Appendix 1.

Safety in river work

Although work at hydrological field stations is not usually dangerous, staff must be careful when working close to rivers that are too deep to wade across, and when working from small boats. There is always a risk of death by drowning.

Fast-moving rivers, especially when in flood, are particularly dangerous. All staff who need to carry out river work must be clearly made aware of the risks involved. Particular risks arise from:

○ wading to carry out current metering when the river is running too deep (or too fast) for this to be done safely

○ carrying out maintenance on staff gauges and water level recorders when a river is flowing fast – especially during flood flows

○ any boat work (it is easy to capsize small boats, even in calm river conditions) – again, especially during flood flows.

All staff who are required to work in or around deep rivers must be able to swim. There is also need for staff to be fully aware of the risks linked to river work. In many countries professional safety and first aid instruction is available for those working on, or close to, deep water. Staff involved should be given appropriate training in rescue and resuscitation techniques.

Travelling through politically sensitive areas

Groups of anti-government activists, and gangs of thieves, sometimes establish bases in the more remote areas of tropical countries. Movement of government employees (including hydrology field teams) through areas where these groups and gangs operate may well be controlled by national security forces. If field teams are allowed to operate in these politically sensitive areas, strict attention must be given to any operational guidelines given by the police or other security forces. Lives and equipment should not be put at risk, however important it may seem to collect hydrological data from these areas.

Emergency radio communication

It is obviously an advantage if field teams travelling and working in remote areas can be in radio contact with headquarters or a regional centre when emergency action is needed. However, remote areas usually have few centres of population, and they are often mountainous – neither of which circumstance makes for easy radio communication. The police and other national security services may have the resources to provide radio communication to their patrols in remote country, but national hydrological services do not usually have the funds to do so.

The best advice is to consult the police, other government departments and parastatal bodies (such as national park and hydropower authorities) and explore what use can be made of the existing communication links which they have in case of emergency. Even if there cannot be a transmitter/receiver in the field team vehicle, it is very useful to know the sites of police posts and other offices where emergency messages can be sent and received.

Dangers from wild animals

Wild animals, however dangerous, usually choose to avoid contact with humans. However, care must be taken where dangers are known to exist, as in rivers where alligators, caymans or crocodiles are active. Hippopotamuses can be dangerous, especially mothers that feel that their young ones are in any way threatened. Hippos on land do not like their routes back to water to be obstructed, so avoid getting in their paths when working on the river bank.

Risks from larger wild animals away from river banks have to be judged with advice from observers and other local residents. The author has worked in a National Park in Africa where the likelihood of meeting buffalo, elephant and other potentially threatening animals was much higher than it would have been in farmland in the same region. Definite rules are difficult to apply, except the need to be on the alert at all times, especially when the team is working outside, and away from, the vehicle. Buffalo have a reputation for being aggressive even when unprovoked, and male elephants (or females who feel that their young are threatened) must be treated with great respect.

Snakes

This is a difficult topic. Local advice may be unreliable, and based more on rumours of dangerous snakes than on actual experience of confronting and dealing with real snakes. It is estimated that worldwide at least 30 000 people die from snake attacks each year, but the statistical risk to a careful field worker is still low. Care must taken if there is a need to walk across ground which is covered by scrub or other dense vegetation, or when disturbing dark, sheltered places in which snakes might choose to rest. It is only really possible to tackle the potential problem of snakes in the context of the local situation. This applies particularly to advice over using anti-venom treatments.

Keep a good lookout also for scorpions and columns of soldier ants. In addition, rivers in marshy and forested areas often provide good homes for leaches.

SECTION 2
DATA COLLECTION

Focus on the measurement of rainfall, streamflow and evaporation

Hydrologists and other specialists involved in water resources planning and management may, depending on their particular needs, require a wide range of field measurements. For example:

○ Water resource planners need nationwide records of rainfall, streamflow and evaporation from which to calculate water balances.

○ Flood prevention engineers need rainfall intensity and streamflow records (taken at frequent intervals during storms and flood events) to study past and present flood water levels and flows.

○ Irrigators, dry land farmers and agronomists need values of rainfall, crop evaporation and soil moisture content.

○ Soil conservationists and geomorphologists need measurements of rainfall intensity and of suspended sediment and bed load in streams.

○ Pollution inspectors need water samples for water quality analysis.

○ Hydrogeologists need well water levels and measurements of the storage properties of rocks in which water can be stored underground (aquifers).

Each hydrological measurement forms a specialist study on its own. One textbook which is concerned only with streamflow measurement has over 500 pages. To keep this book to a manageable size, it is focused on the field measurement of three basic components of the hydrological cycle:

○ rainfall
○ streamflow
○ evaporation.

Measurements of these three components are essential for planning and management water resources in any country. In addition, measurements of any or all of these components are of great value in most of the other areas of hydrological measurement that have just been listed. The three chapters which cover rainfall, streamflow and evaporation are designed to:

○ indicate which measurements are needed
○ describe the instruments most commonly used
○ offer practical guidance on operating instruments.

4
MEASURING RAINFALL

Rainfall and precipitation

Rainfall measurements are essential both for the general management of water resources and for specialized hydrological studies. By strict definition, rainfall is only one form of precipitation, which includes snow, hail, sleet and droplets of water deposited from fog or as dew. In most regions of tropical countries the quantity of precipitation of any other form than rain is very small, except in very high mountainous areas. However, hailstones can fall during intense thunderstorms.

Rainfall events and rainy seasons

To a hydrologist, each separate period of rainfall which contributes enough rain to produce a significant increase the flow in streams and rivers called a rainfall event. In many tropical countries there are distinct periods of each year when the weather is dry, and others when it is wet. The rainy periods may be called 'the rains' or 'the monsoon' in general terms, or they may have particular names in individual countries, such as 'the long and short rains' in East Africa and 'the Northeast and Southwest monsoons' in Sri Lanka.

Raingauges and catches

There are many different designs of raingauge, and in recent years rainfall estimation methods based on radar and satellites have also been used, mainly to estimate rainfall on a regional scale. This chapter concentrates on the types of raingauge that are commonly installed at field sites in tropical countries. These gauges measure the amount of rainwater that passes through a sharp-edged circular metal rim at the top of the gauge. Hydrologists describe the amount of water collected as the catch of the raingauge. Below the rim there is a collector funnel which directs the rainwater into a tube that leads into the body of the raingauge. The rainwater is stored in a container in non-recording gauges, or passes into the measuring system in recording gauges.

Other forms of precipitation

As mentioned above, forms of precipitation other than rain do not normally make major contributions to water balances in tropical countries. However, rim and funnel raingauges will accept hail, sleet and light falls of snow, provided that the catch melts and runs down into the gauge fast enough so that the collector funnel can still receive the continuing precipitation without overflowing. The measured catch in these cases is obviously the liquid water equivalent of the precipitation. Heavy snowfalls require different measuring techniques, which are not discussed here.

Basic raingauge types

The design of a raingauge depends on the purpose for which it is used, but five common types are:

○ non-recording 'daily' raingauges
○ non-recording long-term storage raingauges
○ non-recording simple 'indicator' raingauges
○ mechanical ('clock and chart') recording raingauges
○ electronic recording raingauges.

Fenced daily raingauge site in an irrigated area; the observer's house is to the right (Sri Lanka).

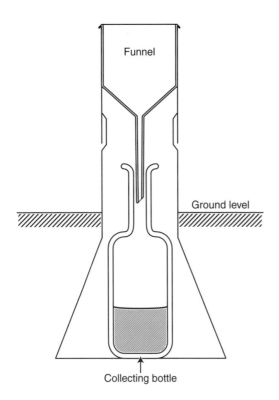

Fig. 2(i) Daily raingauge

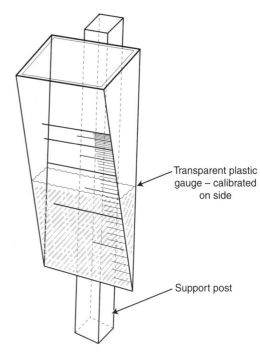

Fig. 2(iii) Simple indicator raingauge

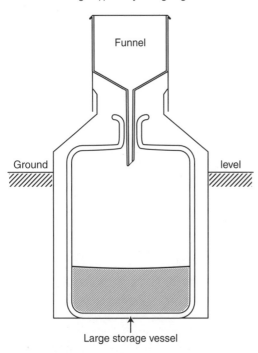

Fig. 2(ii) Long-term storage raingauge

Non-recording 'daily' raingauges

These are by far the most commonly used rain-gauges, and as they are read each day, they are often called daily raingauges. The rainfall catch entering the gauge from the collector funnel is stored in a container, usually a glass bottle. This bottle is emptied every 24 hours, based (in many countries) on a daily reading at 09.00 hours local time.

The British Meteorological Office mark 2 raingauge is widely used in Commonwealth countries. It has a rim diameter of 127 mm (5 inches), which results in the cross-section area of the rim being 126.7 cm^2. The original design has a two-part outer casing made of copper or (more usually in tropical countries) galvanized steel sheet. The upper part of the casing, with rim and funnel, slides closely over the top section of the lower, and must be carefully removed to gain access to the bottle which collects the rainwater. Below the sharp-edged brass rim at the top of the upper casing, the collecting funnel comprises a cylindrical section about 100 mm deep, below which a tapered section directs flow into a pipe which runs down into a glass collecting bottle of about 500 ml capacity. A beer bottle is often ideal, although someone has to drink the beer first!

Within the lower casing, the bottle is contained within a cylindrical can, also of copper or galvanized steel sheet. This can is made watertight – for reasons explained below – and it has a wire handle at the top. After removing the upper casing, the can containing the bottle is carefully lifted out from the gauge and placed on a firm, flat surface. It is essential that the can (with the bottle inside) does not tip over, or the rainwater collected will be spilt and lost before it has been measured. The bottle is then carefully lifted out from the can and the rainwater collected in it is poured into a measuring cylinder.

Rainfall measuring cylinders

These are similar to those used in chemical laboratories, but there is one important difference. The measuring cylinders used in laboratories are graduated in direct units of volume, such as millilitres and litres. Measuring cylinders used with raingauges are graduated so that the volume of water collected indicates directly the depth of rain which has fallen. When the measuring cylinders are produced, a simple equation is used to mark the rainfall depth graduations:

> *Depth of rainfall marked = volume of (the same depth of rainfall × the cross-section area of the raingauge rim)*

Three warnings on the use of rainfall measuring cylinders:

Don't knock them over!

Smaller raingauge measuring cylinders have pointed bases to assist accurate measurement. There is therefore little risk of putting them down before the measurement has been recorded – they will not stand up. Larger cylinders have flat bases, however, and there is a danger of them being placed on the ground and

then accidentally knocked over, spilling their vital contents before the depth of rainfall has been noted and recorded. Keep a firm hold on the measuring cylinder until you have recorded in writing how much rainfall has been collected. Once the rainfall value has been recorded, remember to empty the measuring cylinder.

Emergency use of chemical laboratory measuring cylinders

If stocks of rainfall-graduated measuring cylinders are not available, it is possible to replace a broken cylinder with a volume-graduated measuring cylinder of the type commonly used in chemical laboratories. These laboratory cylinders are more widely available in most countries than rainfall-graduated cylinders, which may have to be specially imported. The use of a volume-graduated cylinder must be marked on rainfall field sheets as well as being noted in the station record book. Local practice will decide whether the conversion from rainwater volume to rainfall depth will be made by the observer on site, or back at headquarters.

Most laboratory measuring cylinders are graduated in millilitres, so that for a British-type raingauge with a five-inch (127 mm) diameter rim, the conversion equation is as follows:

Volume of water equivalent to 1 mm depth of rainfall entering a 5-inch (127 mm) diameter raingauge rim

$$= \pi \times r^2 \times d$$
where r = radius of rim (63.5 mm)
d = depth of rainfall (1 mm)
numerically, the volume is therefore:
$3.1416 \times (63.5)^2 \times 1 = 12\,667$ cubic mm
$= 12.667$ millilitres (ml)

Use the correct cylinder for raingauge rim size

The measuring cylinder used must be the correct one that matches the rim cross-section area of the gauge. Although in most countries only one size of rim will normally be used, keep a check if new types of raingauges with a different rim size are introduced. The risk is that if a measuring cylinder matched to the new rim size gauge is broken, a keen station observer may find that there is still one matched to the old rim size in the station store – and start using it.

If the old cylinder has to be used with the new gauge, details must be noted on rainfall field sheets and in the station record book.

'Throwing back' daily rainfall readings

Of the 24-hour period between 09.00 reading hours on two days, 15 hours come before midnight on the first day (day 1) and only 9 hours in the second day (day 2), but it is at 09.00 hours on day 2 that the amount of rainfall during the previous 24-hour period is recorded.

As the longer time period (15 as against 9 hours) is during day 1, rainfall collected during the 24 hours from 09.00 (day 1) to 09.00 (day 2) is often recorded in data archives against the date of day 1. When that is done, the rainfall total is said to have been *thrown back* from the day of measurement to the previous day.

When using historical rainfall records it is important to check whether or not the rainfall has been thrown back. Field recording forms are often designed so that the measured rainfall total for the previous 24 hours is written on the same line as other measurements taken at 09.00 on the morning of day 2. Rainfall records in the data archives at headquarters may, or may not, have the rainfall thrown back. Always check this point before using historical rainfall records for hydrological purposes.

Coping with heavy rainfall

As already noted, the glass collecting bottle in a daily raingauge usually has a capacity of about 500 ml. Where the raingauge has a rim 127 mm diameter, producing 12.67 ml of water for each mm of rainfall, this size of bottle will accept nearly 40 mm of rainfall without overflowing. To cope with very heavy rainfall, the can which carries the bottle is made watertight so that rainwater overflowing from the bottle can be caught and measured. The need to check these cans for leaks is discussed below.

If the amount of rain caught in the bottle (or bottle plus can) is greater than the capacity of the measuring cylinder, take care when measuring and recording the rainwater:

○ Make sure that both can and measuring cylinder have firm, flat surfaces to stand on.
○ Have a clean, empty bucket, washing up bowl or other watertight container into which the rainwater is poured once it has been measured.
○ Note down each time the cylinder has been filled to its maximum graduated level with rainwater.
○ Note the final reading when the cylinder was only partially filled.
○ Make the final addition, and check the rainwater volume again if there is any doubt over counting the number of full cylinders – then finally empty both the measuring cylinder and the separate container.

An example
If the measuring cylinder has a capacity of 10 mm rainfall, and the catch is 37 mm, you will have noted down that the cylinder was completely filled three times, and then a final reading of 7 mm.

Afternoon readings
In some countries, the raingauge is read – and the collecting bottle is therefore emptied – each day in the afternoon (often at 15.00 hours), as well as in the morning. During periods of heavy rainfall (using the British-type gauge we have been discussing), this will allow up to 40 mm of rainfall to be collected in the bottle, without overflowing, in each of two time periods within 24 hours, namely 09.00 to 15.00 (day 1), and 15.00 (day 1) to 09.00 (day 2).

Make sure that the afternoon reading on day 1 is added to the morning reading on day 2 to give a correct 24 hour total – 09.00 (day 1) to 09.00 (day 2).

Look out for leaks and spiders

As mentioned above, the can containing the raingauge bottle must not leak due to the need

The components of a daily raingauge, from the left: collecting bottle; gauge base sunk in the ground; funnel and upper casing; outer can, with measuring cylinder leaning against it (England).

33

for it to collect rainwater overflowing from the bottle after very heavy rainfall. When checking a gauge in the field always hold the can up to the light. If leaks are detected (as spots of light), the can needs repair. Field teams should always carry spare water-tight cans which can be substituted for any leaking cans – which are then returned to base for repair.

When checking the gauge, also hold the upper (funnel) section of the gauge to the light to check that the pipe is clear. A small, dead spider or a blown piece of twig may allow dust to collect and block the pipe. Blow down the pipe to clear it.

Height of the raingauge rim

The standard height above ground level at which the raingauge rim must be set varies from country to country. British practice, still widely used in Commonwealth countries, is for the rim to be 300 mm (12 inches or 1 foot) above the ground. This was established as a compromise height. It was low enough to avoid serious loss of rainfall catch due to turbulence around the gauge in windy conditions, but high enough to avoid water splashing into the gauge from the surrounding ground surface.

The lowest part of the casing of the Meteorological Office mark 2 gauge is designed to go below ground level. A hole is dug and the gauge installed with the rim exactly horizontal at the correct height above normal ground level. The soil is then packed back around the lower casing section, which has an outward taper to assist stability of the gauge. This has two advantages:

○ The gauge is fixed firmly in position, holding the rim horizontal against disturbance caused by opening the gauge to read it, or by accidental impacts with it.
○ The glass collecting bottle is kept in a cooler position below ground level, reducing evaporation loss on hot days.

Other gauge rim heights
Although the 300 mm raingauge rim height is widely used, there are different national standards for rim heights. Higher levels, of 1.0, 1.2 or 1.5 m, raise the rim away from vegetation and blowing dust nearer ground level. However, experimental results have shown that where the catches from raingauges with rims at different heights are compared on the same field station, those with higher rims tend to collect less rain than those with lower rims.

Non-recording long-term storage raingauges

The highest annual rainfall in many tropical countries occurs in remote, mountain areas where there are few villages to provide safe sites for daily raingauges, and few good roads for easy access by field teams. However, these are often significant water source catchment areas for rivers which are of great importance further downstream. Although it may not be practical to install and operate daily or recording gauges in these remote areas, totals of rainfall collected over periods longer than a day can still be very useful for water resource planning. To collect rainfall over these longer periods, a modified type of non-recording gauge, often called a storage raingauge, can be installed.

A storage raingauge typically consists of a standard raingauge rim and funnel mounted above a storage container of much larger volume than the glass bottle commonly used in a daily raingauge. These gauges may be visited by field teams, or locally-based observers, once every two or four weeks. If there has been a large catch, extra care is needed in counting the number of times the measuring cylinder is filled, and the storage of measured rainwater in a separate container until measurement is correctly completed is especially important.

Some designs of storage gauge are supplied with a cylindrical metal inner container which has a flat bottom and smooth, straight sides. A dipstick can then be used to measure the depth of rainwater in the container, allowing the volume of the catch to be calculated by knowing the cross-section area of container. The dipstick is calibrated like a ruler, with zero on the scale

Locally manufactured long-term storage raingauge. Inner container is some 300 mm diameter by 400 mm deep (Kenya).

Long-term storage raingauge damaged by pastoralists to remove the inner storage can (Kenya).

exactly at the bottom. To measure the depth of rainwater stored, the dipstick is held vertically with its bottom touching the bottom of the storage container. This has an advantage that the container does not have to be emptied on every visit. However, the longer the rainwater remains stored in the container in hot climatic conditions, the more likely it is to evaporate, giving a value for the catch which is less than the amount of rain which actually fell at the gauge site.

Keeping storage gauges safe against damage, by humans or animals, at remote sites can present major challenges. The lower casing should be concreted into the ground and the upper casing firmly fixed to it. A steel bar passing through both sections of the casing (away from the centre rainwater down-tube) can be padlocked to provide a tamper-proof fixing. However, the sharp-edged weapons which pastoral cattle herders carry for defence of themselves and their livestock against wild animals can be used very effectively to cut the upper casing from a storage gauge, releasing the inner container for other uses! The smooth-sided metal sheet containers are specially likely to be stolen. Where the dipstick measurement method is not used, cheap, large plastic water containers (as sold in most rural markets) can be installed.

Wherever possible, encourage the interest of the local people who live and graze their animals in the area, and explain to them why you need to measure the rainfall. This should ensure that there will then be less damage to the storage gauges.

Non-recording simple indicator raingauges

Very simple plastic raingauges, which are often available from irrigation equipment suppliers, can be useful to provide general indications (rather than accurate measurements) of rainfall – but you must be aware of their limitations. They are usually tapered in shape, either as a cone or a flat wedge, and rainfall depth graduation marks are moulded into the plastic walls. These graduations are not as accurate as those on measuring cylinders, and on hot dry days,

after rain, there will be evaporation losses. Although simple gauges are not suitable for national network use, they can help to indicate local rainfall distribution patterns. An example would be across a sugar estate, where a smaller number of 'proper' raingauges would provide sound basic records, whilst a larger network of these simple gauges would indicate distribution patterns. This extra information helps the estate manager to arrange higher or lower levels of irrigation to supplement the rainfall in different parts of estate.

Recording raingauges

These are more elaborate instruments whose records link the amount of rain falling with the time when it fell. They are particularly valuable for hydrology work where the rate of fall (intensity) and precise timing of rainfall must be known, as when studying floods and soil erosion.

The two basic designs of recording raingauge have already been described and compared in Chapter 2 as an example of the use of appropriate field hydrology:

○ mechanical ('clock and chart') designs
○ electronic recording ('tipping bucket') designs.

As detailed descriptions of the way these gauges operate have already been given in Chapter 2, the notes here are focused on particular topics related to gauge operation.

Chart and memory module changing and data down-loading

With mechanical gauges, remember that once the chart has been removed from the clock drum, it can be identified only through the details of station site and period of record written on it when it was removed. The most important details are:

○ name of station site *
○ reference number of raingauge (if appropriate) *
○ date and time of chart installation *

A simple 'indicator' raingauge, as sold for use at home or for small irrigated areas: 'appropriate hydrology' when high accuracy rainfall readings are not essential (Kenya).

- ○ date and time of chart removal
- ○ the amount of rainfall caught in the check gauge (see below) over the period during which the chart was installed.

* This information must be written on the chart at the time it is installed. Make sure that the writing is positioned so that it will not interfere with the likely position of the recording pen trace as the chart is moved forward.

If rainfall records stored in the data logger are to be down-loaded directly into a personal computer (PC), it is essential that not only are full details of the period of record entered into the PC, but also that they are recorded in the field notebook.

Recorder check raingauges

If a recording raingauge is not sited on a met. station, it is normal practice to install a daily raingauge alongside the recording gauge to provide a check gauge. Obviously an observer must be available to read the check gauge each morning. The total rainfall amount indicated by the recording gauge over the period during which a chart (or memory module) is installed should agree closely with the total catch in the check gauge over the same period. If not, the recording gauge almost certainly needs maintenance. Another check is to arrange for the recording gauge to empty the measured catch into a storage vessel mounted below the gauge. The volume of stored water should be equivalent to the rainfall indicated by the recording gauge. Allow for some evaporation losses during hot, dry weather.

Always check a recording raingauge

Whenever you visit a recording raingauge, whether to change a chart or data logger memory module, or because you happen to be at the site for other reasons, always check visually that the instrument appears to operating correctly. When changing a chart, the ink trace will give a good indication as to whether or not the instrument has been operating correctly and the pen writing clearly.

Mark your visit with a 'blip' on the recorder chart

If you visit a station with a mechanical recording raingauge on a day when the chart is not due to be changed (e.g. during the third day of a seven-day chart), open the gauge and move the pen very gently to put a small 'blip' into the trace. Then, again carefully, write on the chart the date and time of your visit, together with your initials, and draw a line to link your writing with the 'blip' you have made on the trace. Avoid writing on a part of the chart where the pen trace may continue before the chart is removed. Check also that the raingauge site name is already written on the chart (if not, add it).

Charts, ink, pens and 'splodges'

The pen and ink system of recording on paper charts is fine when a clean, properly filled pen is writing on the chart during dry weather. However, care is always needed.

- ○ In continuous wet weather, the pen trace may become less sharply defined as the chart paper becomes damp, due to condensation inside the gauge.
- ○ Dirty pens, caked with old, dried-up ink, do not write evenly or continuously.
- ○ If the ink level in the pen is not checked, the pen may dry out.
- ○ Clumsy handling of the pen when installing or removing the chart may leave 'splodges' of ink across the pen trace.

Site visits to electronic tipping bucket recording gauges

Recording the time of a site visit in the electronic memory of a data logger attached to a tipping bucket raingauge should only be carried out after receiving special instructions which are linked to the type of data logging system involved.

Correct exposure of raingauges – beware of growing trees

It is essential to install a raingauge so that its catch correctly represents the rainfall over a wider area around it. If the circulation of air around the gauge site is disturbed by nearby high objects, such as trees and buildings, the catch in the gauge will not truly represent the rainfall of the surrounding area. The word exposure is used to indicate how well the gauge is positioned. The following requirements for good exposure are laid down the Observer's Handbook of the UK Meteorological Office:

> *There should be no steeply sloping ground in the vicinity and the site should not be in a hollow. The site should be well clear of trees, buildings walls or other obstructions. The distance of any such obstacle (including fencing) from the raingauge should not be less than twice the height of the object above the rim of the gauge, and preferably four times the height.*

Beware of growing trees

When selecting a raingauge site, look out for young trees in the area which may grow and overshadow the gauge at a later date. This is specially important if the young trees are on land not under the control of the organization which is planning to install the raingauge. Fast-growing trees, such as eucalypts, can grow up quickly. Make a regular check on the size of trees and bushes round each raingauge site and get them cut back if they are growing too large. Note details of any cutting back in the station record book.

Raingauges in forests – clearings and towers

Measuring rain falling over areas of continuous forest is difficult as the exposure of gauges placed near ground level can be heavily reduced by surrounding trees. Gauges can correctly be sited in forest clearings only if there are no trees in any direction closer to the gauge than twice their height. An alternative is to mount raingauge rims on lightweight metal towers so that the rim level is above the level of the tree canopy. This method is especially use-

ful in coniferous plantations where most trees are of similar height. From the rainguage rim at the top of the tower, a plastic pipe connects to a rainwater storage vessel at the ground level (where any equipment must be installed in a locked container). Operating these gauges may not be easy; pipes can easily become blocked, and may be difficult to clear; theft of parts of the installation may occur. In addition, the height of raingauge rims must be adjusted to maintain them at a 'correct' position in relation to the increasing height of the growing trees. Consult senior hydrologists for detailed advice this topic.

Raingauges operated by national meteorological services

If a raingauge has been installed by the national meteorological service, there is usually no problem for field hydrology teams to visit the site and copy rainfall data. Good working relations normally exist between field inspection teams of hydrological and meteorological services so arrangements to visit each other's raingauges and met. stations are usually easily made. This allows reporting back of any problems at a raingauge or station by whichever field team visits.

A final thought – don't forget school holidays

Schools are often ideal sites for daily raingauges. They provide reasonably secure sites, and both teachers and pupils are often very interested in operating the gauge and using the rainfall data from it in geography and other lessons. However, there can be problems of maintaining good records during school holidays if the teacher who is most keen to collect rainfall data is away from the school site. Before a school term ends, take time to discuss with the teacher concerned, and if necessary with the head teacher of the school, who will be taking raingauge readings during holiday periods.

5
MEASURING STREAMFLOW

Streamflow, hydrometry, stage and discharge

It is useful to start this chapter by defining four important words associated with flow in streams and rivers which are commonly used by hydrologists.

○ *Streamflow* In the English language, a stream is considered to have a smaller flow of water than a river. Small streams combine to form larger rivers. In hydrology, however, the general process of water flowing in channels – of whatever size – is described as streamflow. That word is therefore used in this chapter to describe flow in all sizes (and shapes) of channels. Again for simplicity, the word river will be used to cover all natural channels, whether they could be classed as streams or rivers.

○ *Hydrometry* describes the particular specialist activity, within hydrology as a whole, of measuring flows in rivers and artificial channels.

○ *Stage* When gauging a river, the height of the water surface must be measured in relation to a fixed datum point, usually the zero mark at the bottom of a vertical calibrated scale installed at the gauging station. Hydrologists describe the height of water surface as stage. Stage measurements provide water level information needed to assess flood risk and maintain levels required for navigation traffic; they are also frequently used as indirect indicators of discharge, as discussed below.

○ *Discharge* One of the most important reasons for gauging a river is to discover how much water is flowing past a gauging station in a set period of time. Hydrologists use the word discharge, and a common unit of measurement is cubic metres per second (or 'cumecs' in hydrologists' shorthand). Discharge measurements are needed for many purposes where the quantity of water flowing in a channel must be known, irrigation and hydropower being just two applications.

The challenges of measuring streamflow

Streamflow is more complicated to measure than rainfall. The shape and cross-section area of the rim of a raingauge remain constant, whatever the quantity or intensity of the rainfall. With rivers in natural channels, however:

○ the shape of the channel cross-section through which water flows is different at each measurement site

○ the cross-section area through which the water flows changes as the water surface level moves up or down

○ where the channel is through sand, or other unstable material, the shape of the channel cross-section will change with time

○ the velocity (speed) at which water flows varies at different points across the cross-section of flow

○ measuring flow in deep rivers requires different methods from those used in shallower streams and rivers.

Rivers that do not flow continuously

Before focusing on measurement of rivers where streamflow is continuous, it must be remembered that in dry, semi-arid regions of tropical countries many rivers do not flow continuously throughout the whole year. These rivers can be divided into two general groups:

○ seasonal (or sand) rivers in which flow normally occurs only during rainy seasons
○ rivers which normally flow for much of the year, but in which flow may cease during dry seasons, or under drought conditions.

Precise classification is impossible. Some seasonal rivers flow every year during rainy seasons. Others may flow only once in five years, following intense local storms. As many seasonal rivers have dry, sandy channel beds for most of the year, the name sand river is often used. Methods of estimating flood flows in these seasonal rivers, making use of the position of debris scattered along the banks during the flood, will be discussed below.

Flow or no flow?

Where rivers do not flow continuously it can be important for an observer to note whether water is, or is not, flowing in the channel at a gauging site at a given time, even if it may not be possible to measure the discharge. For national water resource records, this 'flow or no flow' indication is important for periods during which rivers that normally flow have ceased to do so. Gauging stations may be sited along them, for which definite 'zero flow' values can be recorded during periods when no flow is observed.

It will depend on local circumstances whether flows are noted (or measured) in seasonal rivers. Gauging these rivers in which flows are very infrequent – probably for only brief periods as debris-laden floods after storms – is very difficult, as will be discussed below. Seasonal rivers which carry significant flows every year during rainy seasons may have established gauging station sites, in which case observed 'zero flow' readings will form a normal part of long-term records. However, during prolonged dry seasons when there has been no rain, observers and field teams need not travel long distances simply to confirm that such rivers are not flowing.

Why? where? when? – planning streamflow gauging

Having looked at some of the challenges to river gauging in tropical countries, we need to apply the questioning appropriate field hydrology approach before deciding where and how to gauge any river:

○ What is the purpose of gauging?
○ Where does gauging need to be done?
○ When (i.e. how often) is gauging required?
○ How accurately does the gauging need to be done?
○ If an expansion of the regular programme of gauging is needed, what will it cost to set up and to operate?

By asking and answering questions such as these, decisions can be made on which gauging methods to use, what personnel and equipment will be needed, and how much the planned gauging programme is likely to cost.

Gauging stations and sites

When measuring rainfall or estimating evaporation, instruments are normally installed at one place for a long period so the description 'station' – indicating a permanent reading site – can be used. When gauging rivers, however, sites tend to be divided into three types:

○ sites where gauging instrumentation is installed permanently
○ sites where gauging takes place regularly using portable instruments, but where no permanent equipment is installed
○ sites where gauging is carried out using portable equipment for some particular special need, perhaps on only infrequent occasions.

All of these can be called gauging sites, but the first two will be classed as gauging stations,

A stream gauging station. The water level recorder is in the pitched-roof housing at centre. To the right of it is a cableway carriage which allows current metering without wading into the stream. The gauge boards are not visible in this view (Ecuador).

43

especially if they are used to contribute data to a national hydrological archive.

Selection of gauging methods

The choice of streamflow gauging method to use under particular site conditions depends on a number of factors, including:

○ whether flow is continuous or seasonal/irregular only
○ the depth and width of channel at any proposed gauging site
○ the shape and straightness of the reach of the channel (the section of channel immediately upstream and downstream of the proposed gauging site)
○ the stability of the channel – its ability to retain the same shape over time
○ the mean discharge and velocity of flow
○ the variations of stage and discharge between flood flows and low flows
○ the likelihood during floods of the channel being overtopped (flood flows higher than the tops of channel banks) and of damage being caused to measuring structures
○ the likelihood of extensive weed growth in channels
○ the accuracy and frequency of measurement required.

Both stage and discharge measurement methods vary in complexity. Simple methods include measuring river level stage by using upright graduated scales (gauge boards) and estimating flow velocities by measuring the speed of travel of floating objects. At major river gauging stations, however, there will be elaborate, expensively-instrumented permanent gauging installations. The main methods used to measure discharge are listed by R W Herschy in his book *Streamflow Measurement* (for details see Appendix 1) as follows:

○ *Velocity–area method* Velocity measurements made using a current meter are combined with data on the cross-section of the channel through which water is flowing.

When the flow is too fast or too slow to use a current meter, velocity can be measured by recording the time taken for a floating object (a float) to travel a known distance along the channel.

○ *Slope–area method* The discharge is derived from measurements of the slope of the water surface and of the cross-section of the channel over a fairly straight river reach. Numerical coefficients are used to quantify the roughness of the channel boundary surfaces. This method is often used to estimate flood flows, using flood marks, which are indications along the channel bank of the maximum stage during the flood flow period.

○ *Weirs and flumes* To produce more accurate results than those obtained using natural channels, discharge measuring structures of precisely designed shapes can be installed. However, the need for accuracy must be balanced against construction costs and ability to resist damage during flood flows.

○ *Dilution gauging, ultrasonic and electromagnetic methods* These are more specialized methods which are not commonly used in tropical countries.

■ Dilution methods involve injecting a known volume of a tracer liquid into the river at one point and taking water samples at a point lower downstream – the tracer has then been uniformly mixed into the river water. The extent to which the tracer liquid has been diluted by river water can be used to calculate the discharge.

■ The ultrasonic method measures water flow velocity by transmitting ultrasonic pulses diagonally across the channel in two directions simultaneously. Fixed installations are expensive, and the method is not suitable where the water carries high sediment loads.

■ The electromagnetic method measures discharge directly by the installation of a large electrical coil (usually beneath the channel) to generate a magnetic field, then using the moving water to generate an electromotive force. As currently used, this method requires very expensive fixed structure, and is most suitable for stable river channel regimes in temperate countries.

These three specialist methods of measuring discharge will not be discussed further here.

Measuring discharge using the velocity–area method

Of the methods just listed, the one most commonly used in tropical countries is velocity–area, based on velocity measurements obtained using a current meter or floats. In general terms, the quantity of water flowing along a channel during a fixed period of time can be calculated from the following simple equation:

> *Quantity of flow during the time period (discharge)*
> = *velocity of flow* × *cross-section area of flow* × *the time period*

In the real hydrological world, however, the values of velocity and cross-section of flow for this simple equation are not always easy to quantify.

Velocity
The simple discharge equation requires the mean velocity of flow across the cross-section. However, water does not flow along the sides and bottom of any channel (even a smooth concrete-lined irrigation channel) as fast as water nearer to the centre of flow path within the channel. Friction between moving water and channel bed surfaces, such as weed growth or rocks on the beds of natural river channels, causes the flow nearer the channel boundaries to be slower. Most flow velocity measurements obtained from rivers in tropical countries are derived by using current meters, which are described in detail in the next section. A simpler (although much less accurate) alternative method can be used, based on measuring the speed of movement between two fixed points along the channel of floats. The use of floats is also discussed later in this chapter.

Cross-section area of flow
This is calculated from direct measurements (usually made at the time of current meter gauging) of the width of the water surface, and of the depth of the bed below the water surface at fixed distances across the channel. Where the channel shape at a particular gauging station remains stable over a range of discharges and over a long period of time, a steady relationship can be established between stage and the cross-

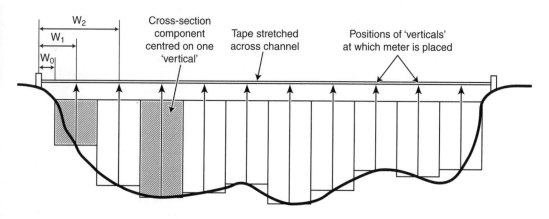

Fig. 3(i) Components of a stream channel cross-section

section area. Where the channel bed is made up of sand and other loose materials, or where the bed is frequently disturbed by flood flows, however, the shape of the bed will be different each time current meter gauging is carried out.

Velocity measurement based on verticals

Before current metering takes place, a measuring tape is stretched across the channel, and a decision made as to how many evenly-spaced positions across the width of the channel (known as verticals) will be selected at which velocity measurements will be taken. Each vertical is the centre line of a segment of the cross-section which reaches half way to the verticals on either side of it. The mean depth of this segment of the cross-section is the distance from the channel bed to the water surface along the line of the vertical.

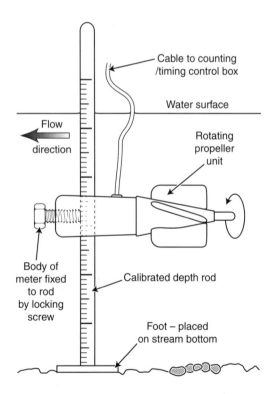

Fig. 3(ii) Propeller-type current meter (in some designs the propeller is replaced by a set of cups rotating around a vertical spindle)

Measuring streamflow velocity using current meters

In tropical countries the most common instrument used to measure streamflow velocity is the propeller current meter, which comprises a small propeller mounted on a streamlined casing which is immersed in the channel, facing directly upstream. The number of rotations of the propeller is transmitted electrically by cable to a separate recording unit, positioned well clear of the flowing water.

In recent years, new designs of electromagnetic current meter have been developed, and these are now being tested in tropical rivers. They rely on the generation of a magnetic field within the body of the meter, and using the moving water to generate an electromotive force, which is related to the velocity of flow at the point where the meter is placed. Although still quite expensive, electromagnetic meters have two particular advantages over propeller meters:

o they can measure very low velocities, where the propeller might stall due to bearing friction and fail to rotate

o they have no moving parts, so they are less at risk from damage in the field than are propeller meters.

Setting up a propeller current meter

At gauging sites where the river channel is shallow – and the flow velocity small – a field team member stands on the channel bed to position the current meter. The meter body is mounted on a calibrated wading rod and positioned using a clamp which allows the meter to be moved up and down the rod, then locked at a fixed position. The rod has a flat 'foot' on its base, which is placed on the channel bed, and is marked in length units, like a ruler. The depth of water is measured using the calibration marks on the rod.

Where the water is shallow enough for a person standing in the stream or river to hold a current meter, the velocity of flow can be estimated from a single reading if the meter is placed at 0.6 of the water depth, measured

Current meter gauging in an irrigation canal (Sri Lanka).

downwards from the surface. For deeper sections, such as those taken from bridges and cableways (see below), readings should be taken at 0.2 and 0.8 of the depth, again measured from the water surface.

Making flow velocity measurements

The following notes apply to using a propeller current meter, with only one (0.6) depth measurement per vertical being made:

a) The measuring tape across the channel is fixed firmly (at both ends) so that it is stretched taut at right angles to the direction of flow with the zero mark on the tape at the bank where measurement is to start.

b) The recording person (on shore) prepares a data recording sheet and checks that the recording unit is functioning correctly.

c) The person in the channel advises the recording person of the width of the water surface at the measuring point.

d) A decision is made as to how many verticals will be used.

e) Knowing the distance from the zero mark on the tape to the nearest edge of the water surface, the recorder calculates the distances along the tape at which the verticals will lie.

f) These distances are noted on the data recording form.

g) The person in the channel then:
   1) wades to the position along the tape of the first vertical
   2) measures the depth of flow at that position (using the meter wading rod) and tells it to the recorder
   3) sets and clamps the meter at the right position on the calibrated rod to read at 0.6 of the depth – measuring from the surface
   4) adjusts the rod foot carefully on the channel bed so that the rod is vertical and the 'nose' of the meter propeller is pointing directly upstream
   5) takes up a position standing downstream of and directly behind the meter, with feet and legs obstructing the flow as little as possible.

6) tells the recorder that the meter is correctly positioned.

h) If current metering is taking place at a staff gauge station, the river level stage must be read and recorded at this time.

i) The recording person then:
   1) checks that a steady signal of propeller rotations is indicated by the recording unit
   2) sets the time period over which rotations of the propeller will be counted (for example, one minute)
   3) starts the recording unit counting
   4) waits until the counter stops at the end of the pre-set time period and records the total of counted propeller revolutions displayed.

The measuring process then proceeds from vertical to vertical until each segment across the whole width of the channel has been covered. Immediately the current metering has been completed, the river level stage must again be recorded from the staff gauge.

Coping with sudden floods

If a sudden flood flow comes through while current metering is in progress, the gauging may have to be abandoned because:

○ staff working in the river may be at risk if gauging continues
○ the flood flows will result in velocities at some verticals being measured at significantly different stages than at others.

Calculating total discharge from current meter gauging data

To calculate the total discharge at the gauged section it is first necessary to calculate the flow through each segment of the overall cross-section which is centred on a vertical. The area of each segment:

○ is centred on a vertical
○ stretches in width half way to the verticals on either side
○ has the mean depth from the water surface to the bed as measured at the vertical.

Current meter gauging. Recorder at left with counting unit and wader at right with meter on wading rod (Sri Lanka).

Current meter gauging from a bridge. A heavy meter is used and gauge boards are attached to the bridge pier (Ecuador).

This area is then combined with the mean velocity through it (obtained from current metering at the vertical) to produce discharge through the subdivision. The total discharge value is then produced by adding together the discharges calculated for each of the segments of the cross-section. The river level stage for which the discharge is calculated is normally the mean of the staff gauge readings at the start and finish of gauging.

Current meter gauging from bridges and cableways

The discussion of current metering has so far assumed that a member of a gauging team can stand safely on the channel bed, holding the meter. With large rivers – or smaller ones in flood – the current meter can be ballasted down by a large weight and lowered into the river from a bridge, or from a steel cable stretched across the river. The cable, together with its supports on the bank and associated equipment, is called a cableway. With some cableways a member of the gauging team rides across the river seated in a small carriage, suspended from the main cable, to position the current meter. With others, the meter can be positioned at a vertical, and raised or lowered by a member of the team standing on one of the channel banks using remote control. As mentioned earlier, on deep river sections current meter measurements will produce more accurate results if readings are taken at 0.2 and 0.8 channel depth instead of just at 0.6 depth.

The need to use the defined sites of bridges and cableways obviously reduces the flexibility in choice of gauging sites. Bridges are usually sited by highway or railway engineers, not by hydrologists. However, they are often placed on straight reaches of rivers to encourage steady flow past the bridge piers and the abutments (the walls on the banks which support each end of the bridge). A bridge sited near a sharp bend in a river channel is at risk of having its piers and abutments scoured out and undermined by swirling water, especially during flood flows. Cableways are expensive, so they are installed

only after careful thought as to whether or not they provide the only effective way of gauging the river at the site concerned.

Safety considerations
Gauging from cableways and bridges can be very dangerous, so safety must always be in the mind of field team leaders and members.

○ When gauging from bridges, take care that one member of the gauging team is directing and controlling vehicle traffic (and inquisitive pedestrians!)

○ Heavy weights are needed to hold the current meter in position in the river – handle them with care.

○ Always be aware of the risks of a member of the gauging team falling into the river, especially when manipulating heavy equipment.

○ Make sure that cableway cables, all supporting structures, and any special support rigs used for gauging from bridges, are regularly checked to the highest safety standards.

Estimating flow velocity using floats

Where it is not possible to carry out a full current meter gauging, a simpler method can be used, based on estimating the flow velocity in the river channel using floats. Two fixed points are selected on the bank along a reasonably straight reach of the river, and the channel cross-sections are measured at each point. Herschy in *Streamflow Measurement* recommends that the upstream and downstream cross-sections should be spaced to allow at least 20 seconds travel time for floats, and that the distance between them should be about four or five times the average width of the river. However, this distance may have to be shortened in small, fast-flowing streams if there is no reasonably straight reach near where measurement of flow is required. It is useful to include a third measuring point midway between the other two. This allows a double check on float velocities.

51

Suitable floating objects are then dropped into the channel upstream of the upper fixed point, ideally at a number of points across the channel (equivalent to current metering verticals). The times of transit between the upper and middle, and middle and lower fixed points are then recorded for each of a number of floats. A mean velocity value is obtained for each 'vertical' by multiplying the surface (float) velocity by a coefficient (typically 0.85). Velocity values for each segment centred on a vertical are then used to produce estimates of discharge.

Simple floats (such as sealed plastic soft-drink bottles partly filled with water, or blocks of wood) are suitable for quick measurements, but more elaborate designs, in which a part of the float travels below the water surface, and which carry coloured 'flags' above the float for easy visibility, can be used. Details are given in the books by Herschy listed in Appendix 1, and in most practical hydrological textbooks. With large rivers in flood, individual tree trunks can be identified upstream, and used as temporary floats as they pass through the measurement reach.

Although discharge measurement based on velocities derived using floats is less accurate than a full current meter gauging, the method can be used:

○ where streamflow is either very fast or very slow, and where current metering may not be successful because
  ■ at high flows it may not be safe for members of the gauging team to wade into the river
  ■ at very low flow velocities a current meter propeller will stall due to friction in its shaft bearings.
○ for rapid appraisal survey work where only approximate discharge values for a number of rivers are required during a very short period of field work.

Stage–discharge relationships

Most regular daily streamflow measurements in tropical countries are of river level stage. These are then related to discharge using stage–discharge relationships, which will now be explored before methods of measuring stage are discussed in detail. At gauging station sites where channels maintain stable shapes, unique numerical relationships between stage and discharge can be developed. The comparatively simple process of measuring stage can then be used to estimate the discharge, which would otherwise require current metering work each time a discharge value was needed.

Stage–discharge, or rating, curves
The stage–discharge relationship for a gauging station is built up from pairs of measured values of stage and discharge. These paired values are usually obtained from current meter gaugings, using the velocity–area method. Gauging must take place at a range of flows and stages, and the pairs of values from each gauging are plotted on a graph. When a number of points have been plotted, spread as far as possible across the expected range of flows and stages, a 'best fit' line is drawn through them, known as the stage–discharge, or rating, curve for the station. Using this rating curve, discharge readings for the station can be obtained by reading the stage from the staff gauge.

The accuracy of the relationship between stage and discharge depends on the number, spread and time frequency of the points which have been plotted. Each new pair of values not only increases the range of flows represented on the rating curve, but also indicates if the stage–discharge relationship is changing.

It is essential to maintain continuous checks on station stage–discharge relationships. Even in rivers with apparently stable channel shapes, minor changes of discharge for any particular stage may occur. As already mentioned, channels with less stable beds need special attention. Pairs of new stage and discharge values must be plotted on the station rating curve at headquarters as soon as possible. An experienced hydrologist must check the stage–discharge plot at regular intervals to confirm the validity of the existing rating curve.

Full details of changes in the station record file

If a new stage–discharge curve has to be introduced (perhaps after a new set of staff gauge boards has been installed), it is essential that the exact date from which it is to operate, and the period over which the previous curve was in use, are recorded on the station record file at headquarters, together with written statements of the equations themselves. When using historical discharge data from the archive records, always check the periods over which different rating curves were in use.

Measuring stage – 1. Staff gauges

Staff gauges and gauge boards

The simplest way of measuring river level stage is to install vertical boards, clearly marked in units of length. The boards themselves are known as gauge boards, and the overall installation (which may comprise a number of boards mounted above each other if the expected stage range is large) is called a staff gauge. Each gauge board is of a precise length, and in a series of boards on a river bank (or fixed to a bridge), the bottom of the scale on the higher board is fixed exactly level with the top of the lower board.

Gauge boards are fixed firmly to steel posts that have been driven into the river bed or bank, or they are attached to fixed structures such as bridge piers. The normal length of each gauge board is two metres. The markings are designed for clear reading by an observer, who may be on the bank some distance away. Markings at every ten centimetres are numbered, with subdivisions (each one centimetre wide) clearly shown by using black and white stripes alternately. Gauge boards can be attached to the sloping wall or gauging structure, but this is not common in tropical countries. With sloping boards, the marked scale has to allow for the angle of slope of the board from the vertical.

It is essential that the zero level at the bottom of the lowest board at a station is below the lowest likely water level at which flow will occur. This is often difficult to achieve when the channel is wide, as the lowest flows may not always travel along the same section of the river bed. Each two-metre gauge board is mounted on a separate steel post, producing a line of boards running up the slope of one bank. The steel post of a free-standing gauge board should have a sloping steel bracing rod added to anchor the vertical post to the bank.

At bridges or other structures, two or more boards may be fixed directly above each other, but they must be attached to the structure so that the zero level on the lowest board is below the lowest likely flow level. The presence of a bridge pier, or other structure, within the normal flow channel will cause a disturbance to water flow. Gauge boards should be mounted where flow disturbance is at a minimum. However, the boards must be sited where the stage value can be read easily by an observer standing on one of the river banks, even under high flow conditions. It is not easy for an observer to read a staff gauge which is fixed to a bridge when standing on the bridge deck itself.

Correct survey levelling essential

Correct survey levelling of staff gauge installations is essential. The zero line on the lowest board is the staff gauge datum, which must be levelled against a station bench mark level on land. That bench mark should be on a permanent building or structure, above likely flood levels, and it must itself be levelled into the national survey network of the country. Full details of all survey work at a station must be entered in the station record file at headquarters, and also in the station record book on site. Precise datum levelling is particularly important when accurate measurements of low flows in a river are needed.

When flows are low, depths of water are usually shallow and any errors in depth measurement will have significant effects on accurate measurement of river level stage.

Damage to staff gauges

During flood flows in tropical rivers, large objects, especially tree trunks, may be carried down. Damage to staff gauges which are fixed into the beds of river channels is likely. After periods of serious flooding, spare gauge board sections should be carried by field teams, who should be ready to carry out repairs to any damaged gauges. If a free-standing staff gauge structure has been seriously damaged or totally carried away during a flood, consideration must be given to whether or not it was originally installed at the most suitable site.

The staff gauge datum of any replacement gauge boards (or complete new installation) must be levelled against the station benchmark. If the datum of the new boards is at a different height to that of those that have been replaced, a new stage–discharge graph must be prepared, together with a fresh station rating curve. Once again, all survey details must be fully recorded in the station record file at headquarters.

Keep the water level staff clean

As the observer will normally stand some distance away from the water level staff when taking readings, it is essential that the gauge board is kept clean so that correct readings of water levels can be taken. Tropical rivers often carry heavy sediment loads, and when river levels fall, gauge boards can easily become coated with sediment. Floods can deposit large amounts of trash around the boards, especially dead or dying vegetation, which also needs removing. Access for cleaning must be considered when the site of the staff gauge is being chosen.

Calendar days and water days

When staff gauges are read daily, the same situation applies as was discussed in Chapter 4 under 'throwing back' daily rainfall readings. Morning staff gauge readings are often taken by observers at 09.00 hours, and the period between these morning readings is sometimes called a water day, as compared with the midnight-to-midnight calender day. The same situation applies as with daily rainfall data, with 15 hours (09.00 to midnight) on day 1 and 9 hours (midnight to 09.00) on day 2.

Once again, care is needed when using archive records of daily streamflows to check whether water days or calender days have been used, and whether or not records have been 'thrown back'.

Older stage records measured in feet

Another point when using historic streamflow records is that in many countries length measurements in feet were used for stage before metric units were widely adopted. These and other pre-metric units are discussed in detail at the end of the chapter. Staff gauge boards were marked in feet which were divided into ten (not into twelve, as inches). These divisions were marked '10', '20', and so on, indicating hundredths of a foot. Archived water level stage records from these earlier times will therefore be in 'feet and hundredths', not in metres and centimetres. A further reminder that inches (12 to a foot) were not used on staff gauge boards.

Measuring stage – 2. Water level recorder gauging stations

A staff gauge provides a measurement of river stage only at the time when it is read. On large rivers during periods when stage changes slowly, one or two stage readings in 24 hours may provide sufficient indications of discharge in the river. However, there are many situations where a continuous record of stage is needed to provide more detailed information to assist analysis of changes in streamflow, especially during periods when flood peaks occur.

Water level (stage) recorders

These are used to provide these continuous records of river level stage. Most recorders used on tropical rivers are operated by a water level float which is attached to a light wire cable (or metal tape), which is in turn wrapped round a drum at the end of a shaft on the

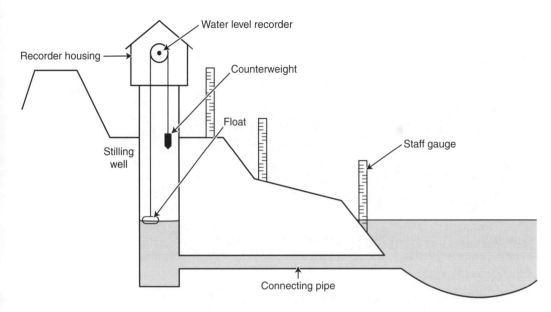

Fig. 4 Water level recorder installation

recorder. The recorder is fixed at the top of a vertical pipe known as a stilling well, which is sited beside the river – to which it is connected by a horizontal pipe below water level. The water surface in the stilling well is therefore at the same level as that of the river. As the stage changes, the water level float moves up and down, turning the shaft on the recorder. Depending on the design, different methods are used to make permanent records of the changes of stage indicated by the turning of the shaft by the float:

○ *Chart recording* Older designs record information by a pen moving on a chart wrapped round a drum, in much the same way as 'clock and chart' recording rain-gauges. The drum may be mounted on the recorder shaft, turned by the rise and fall of the float, with the pen moving parallel to the axis of the drum, or the clock may rotate the drum, with the pen driven from the float by a separate mechanism.

○ *Punched paper tape recording* To avoid the need to translate data from charts, the digital recorder was developed. (The firms of Fischer & Porter and Leupold & Stevens

in the USA were particularly active in developing and marketing this system.) Stage data are logged at regular time intervals by the punching of coded patterns of holes in a paper tape which is moved forward steadily by a clock drive. At the set time interval (e.g. 15 minutes) a punch system is activated which incorporates a line of punch pins, spaced across the width of the paper tape – a 16-track system has commonly been used. A mechanical translation system driven by the recorder shaft alters the punch drive so that each pin does, or does not, punch a hole in the tape at the time interval, depending on the position of the water level float. The different patterns of holes punched at each time interval therefore indicate the stage at that time. At the end of a set reading period the paper tape is replaced, and both old and new tapes must be fully annotated with station, date, time and stage details – in the same way as a paper chart or a memory module from a data logger.

○ *Electronic recording* More modern designs of recorders still rely on a water level float to monitor changes in stage, but data are

55

A clock and chart water level recorder. In this design the clock rotates the drum and changes in water level cause the pen to rise and fall. This recorder was installed by the flume shown on page 66 (India).

collected electronically and stored in data logger memory modules or transmitted directly via telemetry. Telemetry is often used where upstream river levels must be regularly monitored for flood forecasting purposes. The rotating chart drum is replaced by a shaft encoder, which records electronically the position to which the shaft has been rotated by the rise or fall of the float.

Applying appropriate field hydrology thinking, an advantage of digital (punched paper tape) and electronic recording systems is that they collect data which are in digital form and 'computer-compatible'. Charts require the translation of the pen trace although, as already mentioned, electronic chart-readers which convert chart traces into digital form are available. There are also potential problems with pens and ink. However, the chart does allow an observer or field team member to see directly how the recorder is behaving, and how the stage has changed since the chart was installed.

Digital (punched paper tape) recorders offered an earlier alternative technology to charts, but they have now mostly been replaced by electronic recorders. The basic choice currently, therefore, is between the advantages and limitations of chart and electronic water level recorders. The value of chart recorders should not be underestimated.

Recorder installations

When installing a water level recorder, a reach of the river is chosen whose channel characteristics can be expected to maintain a stable stage-discharge relationship. To summarize, a recorder gauging station will comprise:

○ a staff gauge to provide stage readings, and to calibrate and check the stage setting of the water level recorder

○ a vertical tubular stilling well on which the recorder will be mounted, and within which the water level float will rise and fall

○ a horizontal connecting pipe which runs from the lower end of the stilling well out into the river

○ a recorder, mounted on a platform at the top of the stilling well, and enclosed within a lockable housing to prevent unauthorized access

○ a water level float, carried on a wire cable or metal tape which is wound round a drum on the end of the recorder shaft, with a counterweight attached to the opposite end of the cable (on the other side of the drum).

The stilling well must be of sufficient height to keep the recorder housing safely above expected flood levels, and ladder access will be needed if the housing has to be greatly above ground level. The counterweight attached to the opposite end of the float cable provides a downward force on the opposite side of the drum, reducing the apparent weight of the float itself. The float is therefore 'lighter' in the water and more responsive to small changes in stage.

Silting problems

Many tropical rivers carry heavy sediment loads, and connecting pipes to recorder stilling wells frequently become blocked with silt. When installing the station, the horizontal connecting pipe must be of such a size, and mounted in such a position, to discourage blockage by silt and other debris carried in the river or stream. A likely blockage will be indicated by differences between the stage shown by the recorder and that shown by the staff gauge sited beside it. An advantage of a chart recorder is that a poor response of the float to changes in river level stage can be detected directly on site from the shape of the pen trace on the chart. Field teams must always check that the stage indicated by the recorder is the same as that shown by the staff gauge.

When a blockage is suspected, the first approach is to lift the float carefully and secure it close to the top of the well, then to pour (or pump) water into the top of the stilling well. The increased hydraulic pressure in the well is likely to clear the connecting pipe blockage, and the extra water will flush unwanted silt and

Installing a water level recording station. Surveyor at right and survey levelling staff at centre. Two gauge boards have already been installed (Kenya).

other material back into the river. When the float is lowered back into the water within the well, the agreement between the stage indicated by the recorder and by the staff gauge must be checked again.

When stilling wells and connecting pipes have been cleared, notes that the work has been carried out, and of the times between which the recorder was out of use, must be entered on the recorder chart, in the station record book and in the team leader's field notebook.

A removable (and accessible) perforated cap over the end of the pipe in the river may help to reduce blockage problems, but the cap must be removed before the flushing is carried out. If flushing does not clear the blockage, the end of the connecting pipe in the river may be totally covered with silt and other debris. Digging work in the river bed around the end of the pipe may then be the only answer. Road access to a water level recorder station site (at least for four-wheel drive vehicles) for as much of the year as possible is very useful for maintenance purposes.

Wildlife in water level recorders

The enclosed, warm, slightly damp interior of a water level recorder housing can be a popular place for small animals, from frogs, geckos and lizards to various types of insect. They are usually harmless, but be prepared to meet them when you open the housing door for chart changing or inspection. As mentioned in Chapter 3, 'snake stories' can easily become exaggerated, but snakes may choose to rest within water level recorder housings, so take suitable precautions. Remember that the face (and eyes) of the observer will be directly in line with the interior of the housing when the door is opened. Close eye contact with a snake such as an African spitting cobra is definitely not advised! Watch out for swarms of wasps or bees as well.

The best rule is to stand beside the recorder housing on the side where the door is hinged, reach across to unlock the door and swing it carefully towards you. Then move carefully

round to the front of the housing and check – from a distance – whether any wildlife is either leaving the housing, or still resting inside it. Continue to keep a sharp lookout within the housing as the work on the water level recorder continues.

Chart, punched paper tape and memory module changing

Much the same rules apply as for recording raingauges, and no apology is given for repeating notes from Chapter 4. Remember that once the chart or punched paper tape has been removed from station, it can be identified only through the details of station site and period of record written on it. Data logger solid-state memory modules may contain more detailed information stored within them, but they still need to be identified visually. The most important details are:

○ name of station site
○ gauging station reference number
○ date, time and staff gauge stage on installation
○ date, time and staff gauge stage on removal
○ any indication of incorrect operation of the system (recorder and float) during the reading period.

When putting information on to a chart or paper tape during installation, make sure that the writing is positioned so that it will not interfere with the recording pen trace or hole punching pattern. With a memory module it may be appropriate to write the information on a self-adhesive sticky label. An alternative is to note clearly the reference number of the module, together with the necessary details, in a field notebook, making sure that the details from the notebook are correctly copied when the module is sent for processing at headquarters. A similar process applies when down-loading data at site on to a personal computer.

The advantage of the sticky label is that it is attached directly to the module, but beware poor adhesion of labels in wet weather. A strip of transparent adhesive tape stuck over the label after all information has been recorded can save the embarrassment of the field team arriving back at headquarters after a long, wet trip with two separate modules and two separate loose (no longer sticky) labels in the same bag – not knowing which label matches which module!

Charts, ink, pens and 'splodges'

Remember that with the chart, pen and ink recording system, care is needed. The following points, already noted in Chapter 4 on recording raingauges, are repeated here:

○ In continuous wet weather, the pen trace may become less sharply defined as the chart paper becomes damp due to condensation – water level recorder housings tend to be damper places than recording raingauges.
○ Dirty pens, caked with old, dried-up ink, do not write evenly or continuously.
○ If the ink level in the pen is not checked, the pen may dry out.
○ Clumsy handling of the pen when installing or removing the chart may leave 'splodges' of ink across the pen trace.

Always check a water level recorder

As recommended with recording raingauges, always visit and check a water level recorder when you are on site, even if it is not the time to change a chart, paper tape or data logger memory module. Check visually that the instrument appears to operating correctly – a chart will give a good indication of this. Remember with a chart recorder to mark your visit with a 'blip' by carefully moving the recorder shaft to make a slight disturbance to the pen trace and then, equally carefully, writing the date, time, staff gauge stage and your initials in a part of the chart which will not be occupied by the ongoing pen trace (linking those details by a light pen line to your 'blip').

Make sure that the station observer is present during any checking of a water level recorder and discuss with him how the equipment is operating:

○ If the trace on the chart does not indicate a rise in river level after recent heavy rain upstream, has the stilling well connecting pipe filled with silt, or is the water level float leaking?
○ If any flood peaks on a recorder chart seem to be at 'incorrect' times, were the date, time and staff gauge stage entered correctly on the chart when it was installed?
○ How many water level recorder charts/ paper tapes/modules does the observer still have in stock?
○ When was the recorder (and especially the clock with chart recorders) last fully overhauled?
○ When was the recorder pen last cleaned?

As always, details must be recorded in the field notebook and the station record books.

Checking electronic recording water level recorders

As discussed in Chapter 4 on recording raingauges, adding information at the time of a site visit to the memory of a data logger attached to an electronic recording water level recorder should be carried out only after receiving special instructions linked to the type of data logging system involved.

Gas bubbler, pressure bulb and pressure transducer water level recorders

In the section on methods of discharge measurement, dilution gauging and the use of ultrasonic and electromagnetic gauging stations were mentioned as specialized techniques which are not widely used in tropical countries. A similar situation applies with three methods of river level stage recording which have not so far been mentioned:

○ *Gas bubbler recorders* A pipe (or dip tube) runs from the recorder to a fixed point below the minimum expected river level stage. The level of the bottom of the dip tube is precisely related by survey to the stage level datum at the gauging station. An inert gas (usually nitrogen or air) is supplied at a steady rate from a cylinder – like those used to hold gases for oxy-acetylene welding. The pressure against which the gas bubbles out from the bottom of the dip tube varies as the height of the water surface above it. The depth of water above the bottom of the tube can be related to the stage datum for the gauging station, so very precise measurement of the variation of gas pressure can then be used to indicate variations in river level stage.

○ *Pressure bulb recorders* These use a sealed air-filled system. The pressure bulb, which is placed on the river bed at a measured datum level, usually comprises a metal cylinder with a flexible diaphragm across the lower end. A pipe connects the bulb to a pressure measuring sensor and recording system on the river bank. As the river level stage varies the diaphragm moves slightly, and the increase or decrease in air pressure within the system is monitored by the sensor and recorder, often by 'clock and chart' equipment.

○ *Pressure transducer recorders* Various instruments exist which can be used to convert changes in pressure to electronic signals. A summary of types of these pressure transducers which are suitable for river level stage measurement can by found in Herschy's *Streamflow Measurement*.

Although these three more specialized methods of water level recording are not yet widely used in tropical countries, all have the advantage that a stilling well structure is not required. There is a useful comparison of their advantages and disadvantages in the *Hydrologist's Field Manual*, published in New Zealand, details of which appear in Appendix 1. Gas bubblers can operate well in sediment-laden rivers, but they are expensive to buy and complicated to maintain. Pressure bulb systems are useful for short-term studies, or at difficult sites, but they are not as accurate as other systems, particularly where there are large variations in stage at a gauging site. Pressure transducer systems have considerable potential, and in the future they may be developed to provide equipment which is fully reliable for use in tropical rivers.

Gauging flood flows – flood marks and the slope-area method

Sudden flood flows following intense rainstorms present one of the major challenges to gauging tropical rivers. With rivers where staff gauges and water level recorders have been installed, the following gauging problems may occur:

○ If the flood peak passes the station quickly, especially during the night, the local observer may not get to the site in time to read peak stages from the gauge boards.

○ Even if alerted by telephone, field teams will probably not get to the site in time to gauge peak flows.

○ Flood flows will probably overtop the banks of channels, moving beyond flow limits for which stage-discharge rating curves exist.

○ When channel flow is turbulent, the relationships on which the velocity-area method is based may no longer be valid; current meter readings at 0.6 channel depth on the vertical may not correctly represent the mean flow velocity of the cross-section segment.

○ When the flow is very disturbed, debris up to the size of whole trees may be carried along in the flood.

○ There can be serious risks to observers and field teams of being swept away (and possibly drowned) if working close to the river during fast-flowing flood flows.

It is very unlikely that any gauging structures will be installed on seasonal and sand rivers, where flows may only occur after intense tropical storms.

Flood flow overtopping a small gauging station site. Although the top of the gauge board is still visible (centre right), flow is no longer confined to a properly defined channel (Kenya).

Flood marks

Although it is very difficult to measure peak flood flows at the times when they pass a station, they leave behind them lines of debris along the banks whose position can be useful to hydrologists. These lines of debris, known as flood (or rack) marks, are deposited as the stage begins to fall after the peak of a flood, and they provide some indication of the upper limits of the cross-section of flow at the time of the flood peak.

The slope–area method of estimating discharge

To provide some indication of otherwise ungauged flood flows, the cross-sections obtained from the flood marks may be combined with estimates of:

○ the slope of the channel
○ the 'roughness' of the channel bed.

The slope–area method

This was originally developed to estimate discharge in ungauged channels where flow was less turbulent than during peak floods. It is described in detail in textbooks such as those by Herschy which are listed in Appendix 1. One of the most widely used equations is that by Manning (see Appendix 3), which relates discharge to values linked to the channel cross-section and slope. A roughness coefficient ('n') is used which is itself linked to the resistance to flow offered by the channel bed (e.g. by large rocks or extensive weed growth). During flood flows the ideal conditions to satisfy the Manning equation do not apply, but simplified slope–area methods such as those described in *Hydrometry, Principles and Practices* (edited by Herschy) can be used to provide rough estimates of flood flows from the position of flood marks.

Discharge measuring structures – weirs and flumes

Most stage and discharge values obtained in tropical countries come from gauging sites which use natural river channel cross-sections.

However, more accurate measurements can be produced by using specially designed flow measuring structures built into the channel bed, which can either be weirs or flumes. In both cases the design incorporates a standard stage–discharge rating curve, and if the structure is built and installed correctly the operational rating curve should be very close to the design curve. The structure will then offer a rating curve which:

○ is considerably more accurate than one for a natural channel
○ will not change with time (assuming the structure is correctly maintained).

Although weirs and flumes are widely used to measure irrigation water flows in tropical countries, especially for gauging small artificial distribution channels, their use on natural river channels for more general water resources assessment is much more limited. They will therefore be mentioned here, but not discussed in detail. Reasons why weirs and flumes are not used widely for gauging tropical rivers include:

○ the difficulty of designing single structures to measure a wide range of different flows to the same level of accuracy
○ high sediment loads in tropical rivers resulting in heavy silt deposits upstream of gauging structures, preventing them from operating correctly
○ damage to structures by trees and other floating debris during flood flows
○ the high cost of structures, especially on large rivers.

The simplest form of weir (not necessarily used for gauging) is a wall or embankment across a river, often built to maintain a higher level of water upstream than would occur with the natural channel. If the top of the weir is truly level, however, water will flow over it at an equal depth right across the structure, and mathematical equations exist which link the depth of water flowing over a weir and discharge. With all weirs a staff gauge upstream of the top (or crest) is required (and possibly a water level recorder) to measure accurately the

A compound weir structure. Thin-plate weir sections are used at the centre of the structure (Kenya).

depth of water flowing over the weir. Weirs are much more commonly used than flumes for gauging rivers.

Types of measuring structure

In *Streamflow Measurement*, Herschy divides measuring structures into five categories:

○ thin plate weirs
○ triangular profile weirs
○ flat V weirs
○ broad crested weirs
○ flumes.

○ *Thin plate weirs* These tend to be used for small streams or irrigation channels. The measuring sections are usually made of steel sheet, although the lower bulk of the structure may be of concrete. The shape of the measuring section may be a rectangular notch or a 'V' shaped notch, through which water flows, or a straight crest, over which water flows. Equations exist to relate stage and discharge for standard shapes of thin plate weirs. Note that these weirs behave differently if the sides of the plate are at some distance from the edges of the channel water surface (contracted weirs) than if the sides coincide with the edge of the channel (suppressed weirs). The discharge for the same depth of water passing through or over the same shape of thin plate will differ if the weir is contracted or suppressed. Tables relating stage and discharge for thin plate weirs appear in Appendix 3.

○ *Triangular profile Crump and flat V weirs* These are designed specifically for accurate streamflow gauging, but are not widely used in tropical countries. They are triangular in cross-section (looking at right angles to the direction of flow), so that flow close to the channel bed has to rise to pass over the crest of the weir. Crump triangular profile weirs have crests which are level, whilst flat V weirs, as their name implies, have crests which gently slope down towards the centre line of flow across the structure. Flat V weirs are preferred to Crumps when low

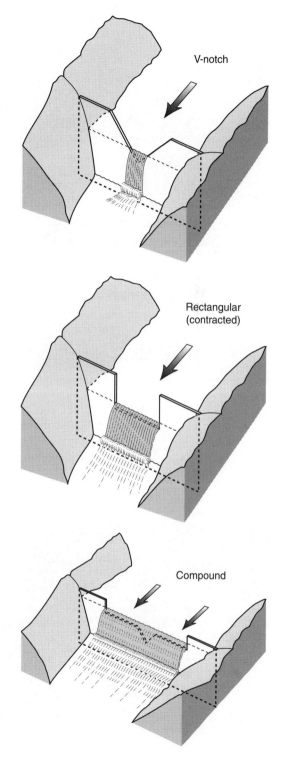

Fig. 5 Thin plate weir designs

65

A small flume installed to measure surface runoff from soil conservation trial plots. Similar flumes are used for small-scale irrigation flow measurements (India).

flows across the structure are likely to occur regularly.

○ *Broad crested weirs* These have crests which, when seen at right angles to the direction of flow, are much wider than with thin plate or triangular profile weirs. Although they can be built as very robust structures, large enough to handle significant river flows, a carefully levelled crest (designed for accurate gauging) may be damaged or partly fouled by large debris during flood flows.

○ *Flumes* These are structures with specially shaped channels through which flow takes place. For each flume design a standard relationship exists between stage and discharge. By far the widest use of flumes in tropical countries is to measure flow in small irrigation channels, and details can be found in specialized textbooks such as the *Water Measurement Manual* and *Flow Measuring Flumes for Open Channel Systems*, detailed in Appendix 1.

Compound weirs

Compound weirs are often used where flows at a station vary considerably between wet and dry seasons. They combine different weir sections in the same structure. There is usually a lower central section (often a V-notch or rectangular thin plate, or a Crump) to gauge low flows at the centre of the structure, with straight crest sections (either rectangular thin plate, Crump or broad crested) on either side of it to accommodate larger flows.

Sources of more detailed information on hydrometry

Following the overall approach adopted for this book, the aim of this chapter has been to give a broad overview of the various techniques and instruments which are commonly used for streamflow measurement in tropical countries. For more detailed information on hydrometry, two recommended sources (which have already been mentioned) are the World Meteorological Organization *Guide to Hydrological Practices*, and *Streamflow Measurement* by the British hydrometric expert R W Herschy.

The International Standards on Liquid Flow Measurements in Open Channels have also been mentioned. Further details of the Standards, and of the International Organization for Standardisation (ISO) which publishes them, appear in Appendix 2.

Safety in river work

This very important topic has already been discussed in Chapter 3, and a number of additional safety matters have already been discussed in this chapter. A reminder that particular risks arise from:

○ wading to carry out current metering when the river is running too deep (or too fast) for this to be done safely

○ carrying out maintenance on staff gauges and water level recorders when a river is flowing fast – especially during flood flows

○ any boat work (it is easy to capsize small boats even in calm river conditions) – again, especially during flood flows.

Remember too that:

○ All field staff must be made fully aware of the risks linked to river work.

○ All field staff who are required to work in or around rivers should be able to swim.

○ Safety life jackets should be worn at all times where rivers are running deeper than shallow wading depth – below knee level.

○ Professional safety and first aid instruction, including training in rescue and resuscitation techniques, should be arranged as appropriate.

Dangerous wildlife in rivers

Another reminder from Chapter 3 is that wild animals normally choose to avoid contact with humans. However, be careful when working in or beside rivers where alligators, cayman or crocodiles are known to be active. In Africa, be wary of hippopotamuses, especially mothers who feel that their young ones are in any way threatened or adults who feel that their route from land back to the water is being obstructed.

Flow measurement unit confusion – cumecs, cusecs and mgd

This is a final section on flow measurement and water volume units, on which more details appear in Appendix 3. Hydrologists and water engineers, like many other specialists, have their own 'shorthand' notation which can be confusing to newcomers. The use of the short-hand term 'cumecs' to indicate discharge in cubic metres per second has already been mentioned. Another area of difficulty for newcomers is the use of non-metric measurement units, especially in historical hydrological records.

Engineering shorthand notation
Scientists, when expressing a flow in cubic metres (m^3) per second (s) use expressions such as m^3/s or $m^3 s^{-1}$. Practical water engineers, however, have developed their own shorthand notation which is widely used, although it is not as precisely 'correct' as the scientific expression of units. Examples include:

- *cumec*   cubic metre per second, commonly used to express discharge
- *cusec*   cubic foot per second, the Imperial units predecessor of cumecs – the American version is second-feet
- *mgd*   million gallons per day, popular amongst water supply engineers (see note below on Imperial and US gallons)
- *tcmd*   thousand cubic metres per day, the metric successor of mgd
- *Ml/d*   megalitres per day, a (numerically equal) alternative to tcmd
- *acre-foot* one foot depth of water across the area of one acre, a volume measure especially popular with American engineers (who also use acre-inch)

- *Ha mm*   one millimetre depth across the area of one hectare.

Non-metric units
The metric SI measurement system, with the metre as the basic unit of length, has been officially adopted by governments in many countries, and should be taken as a standard for the future. However, other systems are still in use. Many countries have their own traditional units for measuring land area, such as the *feddan* in the Sudan, which are still used, especially by older people in rural areas and in legal documents related to land holding. These land area units may still be used in areas where irrigation has been long established.

More widely used has been the British Imperial measurement system, based on yards, feet and inches for linear measurement, acres for land area and gallons for quantity of water. These Imperial units may still be used by older people, and the use of feet on staff gauge boards in pre-metric days has already been discussed. Some conversion factors between the standard metric and Imperial units of particular relevance to field hydrology are listed in Appendix 3, but conversion tables are widely available for basic units such as those of length and volume.

Using 'mgd' – a warning on gallons
When working with older textbooks, manuals or project reports, remember that the gallon is not the same size in American US and British Imperial systems of units:

> *1 British Imperial gallon = 4.55 litres;*
> *1 US Gallon = 3.78 litres;*
> *therefore 1 US gallon = 0.83 Imperial gallons*
> *and 1 Imperial gallon = 1.20 US gallons.*

When working with discharges in 'million gallons per day (mgd)', always be sure to check whether Imperial or US gallons are being used.

6
ESTIMATING EVAPORATION

Evaporation – a complex process

Evaporation, the third major component of the surface water balance, is the movement of water from land and water surfaces into the atmosphere. As water evaporates from a surface, it changes from liquid form into water vapour, which then mixes with the air moving above the ground. Accurate measurement of the rate at which evaporation takes place over land and water surfaces is therefore more difficult than measuring rainfall and streamflow, both of which involve water in liquid form acting under the influence of gravity.

As evaporation can be a complicated component of hydrological water balances, more space is devoted in this chapter to explaining the processes of water movement involved, and the need to measure evaporation, than has been the case in the chapters on rainfall and streamflow.

Two very important factors controlling evaporation rates from land and water surfaces are:

○ the ability of the air above an evaporating surface to absorb and carry away water vapour
○ the supply of water to the evaporating surface.

A wind blowing across a puddle in a road after rain causes the water to evaporate. However, when the puddle dries up, evaporation stops – there is no water left to evaporate.

The ability of the air to absorb water vapour

Most of this chapter is focused on methods of quantifying the ability of the atmosphere to take up water vapour produced by evaporation. A range of climatic elements is involved, especially:

○ air temperature
○ air humidity
○ solar radiation (related to intensity of sunshine)
○ wind speed.

Returning to the puddle, it will evaporate faster when the air is hot and dry, when the sun is shining, and when the wind is blowing more strongly.

Is there any water to evaporate?

The second major factor which controls evaporation is the supply of water to the evaporating surface. Consider two practical examples:

○ A hot, dry wind blowing across a desert and then across a lake, causes water to evaporate from the lake, but not from the desert.
○ A hot, dry wind blowing across farmland – first across a field with a dry, bare soil surface and then across a field where a growing crop is irrigated – causes water to evaporate from the irrigated field, but not from the dry soil surface.

The first example highlights the difference between evaporation from water and land surfaces. The rate of evaporation per unit of surface area of a lake will largely be controlled by atmospheric conditions, especially the ability of the air to take up water vapour. However, the total quantity of water evaporated from the lake will also depend on its surface area.

Evaporation from land surfaces is more complicated. The supply of water to evaporate varies according to the type of surface. In the case of the desert, no water means no evaporation. The dry soil surface in the second example is a more complex case than desert sands. After rain has wetted a bare soil surface where there is no cover of vegetation (either as living plants or as dead plant remains), a hot, dry wind will encourage evaporation from the wet soil surface. The upper layers of the soil will dry out quickly, but upward movement of water through the structure of the soil is slow, so the deeper levels in the soil profile may still hold water.

Transpiration

Plants use their roots to extract water stored in the soil. The food that plants need is dissolved in that soil water in the form of chemical solutions. To maintain the flow of these food-bearing solutions from the soil, and to distribute them within the plant, there must be a steady flow of water through the structure of the plant from roots to leaves. To maintain this flow, the plant allows water to evaporate into the atmosphere through very small holes in the leaf surfaces called stomata or stomates. As water evaporates from the stomata, more is drawn up through the roots and through conducting passages within the plant, carrying with it plant food in solution. This process is called transpiration, and the plant can control it by closing the stomata on the leaves when it is losing water through them faster than it can absorb water through the roots.

Returning to the second example, when a hot, dry wind blows across the irrigated crop after rainfall, water will be available to evaporate from:

○ the stomata on the crop leaves – supplied via the roots from soil layers wetted by irrigation and rainfall
○ plant surfaces which have been wetted by rainfall, or by sprinkler irrigation
○ soil surfaces wetted by rainfall and irrigation.

'Evaporation' or 'evapo-transpiration'?

Some specialists in soil–plant–water relationships prefer to use the word 'evapo-transpira-tion' to describe this combined loss of water by evaporation from the leaf stomata, and from plant and soil surfaces. The author's preference is to use 'evaporation', and to provide more details (where necessary) of the surfaces from which evaporation is taking place.

Specialist evaporation calculations

The practical field measurement methods which will be described below produce numerical values of the mean rates of evaporation across the area around the field station where data are collected. Where values are needed of evaporation from a particular lake or forested area, from an irrigated area under a particular crop – or even from a single tree – extra stages of calculation are needed, and extra measurements usually need to be made. Specialist evaporation measurements are not discussed in detail in this chapter, although the use of crop factors to estimate evaporation rates for the different stages of growth of agricultural crops will be discussed later, as will soil moisture measurement.

Why measure evaporation?

If evaporation is so difficult to measure, why do hydrologists, water supply engineers, irrigators and water resource planners need numerical values for it? Because it is, together with rainfall and streamflow, a major component of surface water balances. Evaporation values are also used in water balance studies involving drainage through soil profiles and the recharge of groundwater storage.

Water balances are not just of scientific hydrological interest. They are an essential part of the commercial operation of major water resources projects where water, so often regarded as 'free', has to be given a financial value. Evaporation is a major component of water balances of such projects in hot, dry areas, as illustrated by the following two cases:

○ the efficient operation of storage reservoirs for water supply and hydro-electric power generation
○ the effective irrigation of farm land.

The financial value of water evaporated from reservoirs

Large storage reservoirs for water supply or hydro-electric purposes are very expensive to establish (due to the high costs of building dams) and to operate and maintain. If the organization that built the dam is required to be self-financing, costs must be regained by charging customers for water supplied or electricity generated. All water stored in the reservoir which can be extracted (whether for water supply or hydro-electric generation) has a financial value for the operators of the dam. The dam operator therefore considers water lost from the reservoir surface by evaporation as 'money lost'.

The financial value of water for irrigation

Irrigation is the process of replacing water which has been evaporated from farm crops and the soil in which they grow, or which has drained down below the crop root zone. Moving water to irrigate farm land can be expensive. Where the water flows in canals and channels, complex sluice structures may be needed to control water flow, and channels may have to be lined with concrete to prevent leakage losses. Pumps may be required to lift water to levels where it can flow along channels by gravity, or to operate sprinkler or drip irrigation systems. Pumps are not only costly to buy and maintain, but they also require diesel fuel or electricity to operate them. So, like the water stored in reservoirs, irrigation water has a financial value. The more accurate the farmer's knowledge of how much water irrigated crops are evaporating, the more efficiently he or she can provide that water, when and where it is needed.

Water has been stored in reservoirs and used for irrigation for many centuries in many countries. However, over the last 200 years there have been pressures to build larger storage reservoirs and irrigation schemes, and to operate them to use water as efficiently as possible. Faced with these pressures, hydrologists and water resource engineers have worked to produce ever more accurate numerical values of evaporation for use in water balance calculations.

Measuring or estimating evaporation?

Faced with the complexities of measuring evaporation directly as a movement of water vapour in the atmosphere, research scientists and practical hydrologists explored how it could be estimated indirectly by taking simpler measurements. Three broad approaches resulted from those studies:

○ measuring direct evaporation of water stored in instruments known as evaporimeters, of which the evaporation pan is most widely used

○ developing mathematical equations to estimate rates of evaporation from measured climatic factors – air temperature and humidity, sunshine and wind have already been mentioned

○ developing specialist instruments to measure evaporation as the movement (or flux) of water vapour carried in the air away from land and water surfaces.

The first two approaches, which have been widely used for practical water resource management in tropical countries, will be discussed in more detail below. The third approach has so far been largely confined to research experiments and will not be explored here.

Rates of evaporation – units and time periods

The rates of evaporation over land and water surfaces are normally expressed (in the same way as rainfall) as depth of water per unit area of land and water surface in a defined period of time. The most common expression is in millimetres per day.

'Calendar' and 'reading' days

As instruments at met. stations are normally read at a standard time (often 09.00 hours) each morning, the data from which daily estimates of evaporation are prepared indicate the climate during the 24 hour period from 09.00 on the previous day, not the midnight-to-midnight calendar day. The idea of data being 'thrown back', discussed in relation to rainfall in Chapter 4, also applies to climate measurements for evaporation. For example, the readings recorded at 09.00 on 23 March will be used to produce an evaporation estimate for the previous day, 22 March, as 15 hours of the 24 since 09.00 were before midnight on the calendar day of 22 March.

Open water, potential and actual evaporation

Three evaporation processes are most important to hydrologists, water resource planners and irrigators:

○ *Open water evaporation* takes place from the surface of any exposed body of water such as a lake or reservoir.

○ *Potential* and *actual evaporation* These need to be explained together. Returning to the twin controls of evaporation – the ability of the atmosphere to absorb water, and the supply of water to a surface to evaporate – consider two fields of grass, one of which is irrigated, the other not. Rain has fallen, after which a hot, dry wind starts to blow. What happens to the evaporation rates?

■ The irrigated field, with the grass roots well supplied with water, continues to evaporate at a rate linked to the ability of the atmosphere to take up water. Hydrologists classify the rate of evaporation from a short crop (such as grass) which is growing well, and is plentifully supplied with water, as potential evaporation. The word 'potential' indicates that the rate of evaporation is close to the highest level which the climatic factors can generate from short vegetation. Experiments have shown that potential evaporation rates from well-watered uniform short vegetation are approximately 0.7 to 0.8 of open water evaporation rates under the same atmospheric conditions.

■ The evaporation rate from the non-irrigated field falls as the water stored in the soil profile is used up. The hot, dry atmosphere will encourage high rates of evaporation through the stomata on the leaf surfaces. If this is not controlled, however, the plants will first wilt, and may eventually die if the hot, dry weather continues. The plants therefore conserve their water status by reducing the openings of the stomata on their leaves. Under these conditions evaporation cannot take place at the potential rate, but at a lower rate described as actual evaporation.

Crop factors

Before leaving potential and actual evaporation, it is worth noting that a crop evaporates different amounts of water depending on its stage of growth. Imagine two separate days during the life of a maize crop when the weather and climate conditions are the same:

○ a day when the seedling plants have just emerged from the soil
○ a day when the vegetation is full and green.

The young emerging maize seedlings will extract much less water from the soil than the crop when fully grown. To match the use of water by a crop to its stage of development, potential evaporation rates can be multiplied by a crop factor. Values of these factors for a range of different crops are listed in the FAO Irrigation and Drainage Paper No.24, which is discussed in the next section below.

Important publications on evaporation from FAO

A very useful, and widely available, group of publications covering practical aspects of the use of evaporation estimates in agriculture is the *Irrigation and Drainage Paper* series published by FAO, the Food and Agriculture Organization of the United Nations.

An excellent introduction to the practical use of evaporation measurements in managing irrigated farming is *Guidelines for Predicting Crop Water Requirements* by J Doorenbos and W O Pruitt (Irrigation and Drainage Paper No. 24). The evaporation estimation methods recommended in it will be discussed more fully below. A full reference appears in Appendix 1, but to practical irrigators it is known simply as *FAO 24*. Being an FAO publication, it is available in most tropical countries.

CROPWAT

A related FAO Irrigation and Drainage Paper is No. 46, *CROPWAT – A Computer Program for Irrigation Planning and Management*, which provides computer programs on disk (and associated guidance) using the evaporation and irrigation water requirement calculation methods presented in *FAO 24*.

Agro-meteorological Field Stations

An FAO Irrigation and Drainage Paper which is especially useful when planning field stations to estimate evaporation is No. 27 *Agro-meteorological Field Stations* by J Doorenbos. This provides a very good introduction to the design and operation of met. stations where climatic measurements are made from which evaporation can be calculated. The first chapter in Paper No. 27, on basic requirements, covers the selection of suitable sites and instruments, and the arrangement of observation procedures. Chapters then follow on each of the following topics:

o   temperature
o   humidity
o   wind
o   sunshine and radiation
o   precipitation
o   evaporation
o   interpretation and use of data.

As an FAO publication, Paper No. 27 is widely available, and is recommended reading for field hydrologists who are involved with collection of climatic data from met. stations.

Practical field methods for estimating evaporation

Two broad approaches to estimating evaporation under field conditions have already been mentioned:

o   measuring rates of evaporation of water stored within evaporimeters
o   measuring climatic factors and using the data in mathematical evaporation equations.

The most widely used evaporimeters are evaporation pans. These can be sited at met. stations where other climate factors are measured, or installed at sites where the only other instruments are raingauges.

The second approach requires measurement of a range of climatic measurements. One of more complex (and accurate) evaporation equations, developed by Dr H L Penman at the Rothamsted agricultural research station in England, requires data on air temperature and humidity, solar radiation and wind speed. Such measurements are normally made at a met. station which has been established by the national hydrological or meteorological organization.

Commercial organizations with a particular interest in evaporation data, such as hydro-power authorities or irrigated sugar estates, often also set up their own evaporimeters and met. stations to provided detailed information on the climate within the area of their activities. The climate data collected at these stations are normally made available for use by the national organizations.

Evaporation pans

The instrument most commonly used world-wide to estimate evaporation is the evaporation pan, which is a small, open tank containing water. The amount of water lost from the surface area of the tank in a given time (usually 24 hours – morning to morning) gives a rate of

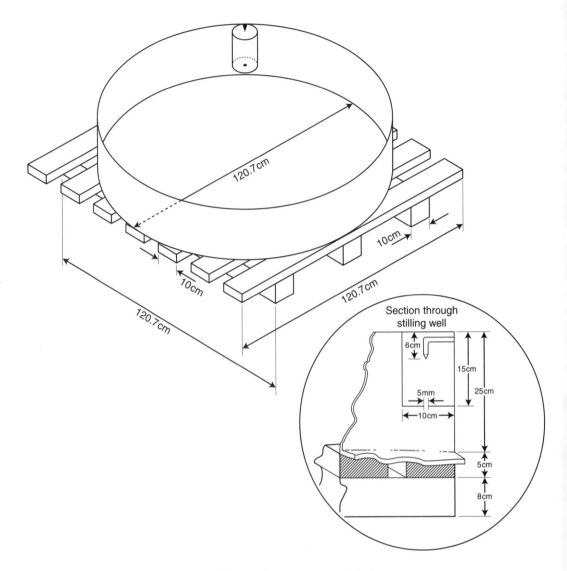

Fig. 6 Evaporation pan: Class A design

evaporation. Allowance is made for addition of water to the tank by rainfall. One great advantage of the evaporation pan is that it is simple to operate and cheap to install.

However, applying the appropriate field hydrology approach, the simplicity and cheapness of pans has to be balanced against the accuracy of the evaporation estimates which they produce. Like other types of evaporimeter, a pan provides a direct measurement only of the amount of water lost from the instrument itself. The rate of loss of water from a pan is closer to evaporation rates from the land around it when plant and soil surfaces are wet than when they are dry. Experimental results have shown clearly that more accurate estimates of evaporation can be obtained using evaporation equations based on a range of climatic measurements made at a met. station. However, as will be discussed below, met. stations are much more expensive to equip and operate than evaporation pans.

An unscreened Class A evaporation pan. It has a good, open site (Mauritius). (Photograph by David Kirby)

In the 1950s and 1960s there was a great deal of technical discussion over the shape and size of evaporation pans, and over whether or not they should be sunk into the ground. The result of these discussions was a general agreement that the United States Weather Bureau 'Class A' pan design was an acceptable international standard. The Class A pan is an open-topped circular tank, with internal dimensions 0.254 m (10 inches) deep and 1.207 m (47.5 inches) diameter. Although ideally these pans are made from an expensive alloy known as monel metal, in tropical countries they are usually welded up from galvanized steel sheet. Depending on the national practice adopted in any country:

o The outside of pans may be painted (often with white or aluminium paint), or left with the galvanized finish.
o The inside of pans may be painted with black bitumen paint, or left as unpainted galvanized sheet.
o A light wire-mesh screen (often of chicken wire mesh) may be fitted over them to prevent birds and animals from drinking the water.

The pan is mounted above ground level on a wooden framework made of timber at least 50 mm by 100 mm, treated with a wood preservative to discourage rot. For the pan to operate successfully, it must be mounted exactly level and the water surface must be maintained at a distance of between 50 and 75 mm below the top of the pan. There are two basic methods of operating a Class A pan:

o *Measuring cup method* An indicator which shows a fixed water level is attached to the wall of the pan, and water is measured into, or out of, the pan each day (using a special calibrated cup) to return the water surface to exactly the set water level.
o *Hook gauge method* An adjustable depth gauge is placed on a specially levelled section at the side of the pan each day, and a precise water level measurement is taken.

Measuring cup method
The indicator fixed to the pan wall is normally an upwardly pointed metal rod, and the precise water level is shown when the sharp point of the rod just touches the water surface. A curved shield protects the area around the point from disturbance by ripples across the pan water surface produced by wind, but openings below the water table allow water movement from the main part of the pan. A special measuring cup is used, whose volume is the pan surface area multiplied by one millimetre depth. For a Class A pan, the volume of the measuring cup will be:

Cross sectional area of the pan multiplied by one millimetre depth:

$$= \pi \times r^2 \times d$$

where r = radius of pan (603.5 mm)
d = depth of rainfall (1 mm)
numerically, the volume required for the cup is therefore:

$$3.1416 \times (603.5)^2 \times 1 = 1,144,209 \text{ cubic mm}$$
$$= 1.144 \text{ litres}$$

As when measuring rainfall, the observer must be very careful when counting the number of full cups of water, and note down the number straight away. Cups of water are:

o *added* to the pan to bring the water level to the fixed point when evaporation over the past 24 hours has been more than the rainfall
o *removed* from the pan when rainfall has been more than evaporation.

Hook gauge method
An adjustable depth gauge also uses an upwardly pointed rod to indicate precisely the water surface level. To shield the gauge point area from ripple effects, an upright metal tube called a stilling well is used, the top of which is set precisely level by using three levelling screws around the base. The main body of the hook gauge (spider) has three projecting arms which rest on the levelled top of the stilling well when readings are being made. The pointed rod which indicates the water surface level curves upwards from the bottom of a central

A hook gauge on its stilling well in a Class A evaporation pan. The pan is painted white and the levelling screws on the stilling well can be seen (Saint Helena).

measuring rod, which is itself marked with millimetre graduations and has a screw thread. This central rod passes through a rotating threaded collar, which is attached to the main body of the gauge. To read the water level, the collar is turned, moving the measuring rod up or down until the point just touches the water surface. The distance measured along the central rod then indicates the water surface level.

When the water level rises above 50 mm below the top of the pan wall, or falls below 75 mm, water is removed or added to bring the level nearer to the 60–65 mm level, and a fresh water level reading is taken. As when measuring water in and out of the pan using the cup method, full attention from the observer is needed when using the hook gauge, especially when water has to be removed from, or added to, the pan.

Screening pans against birds and animals

In dry areas of tropical countries, an open container of water is an obvious attraction to the birds and animals searching for a drink. This drinking increases the apparent evaporation loss from the pan, especially if the animal is an elephant! However, it is normally birds and small animals which cause the drinking problem and a simple answer is to fit a screen over the pan surface to prevent the animals reaching the water surface. Two important points to remember about screening pans:

○ It must be clearly noted on all records whether or not the pan is screened.
○ A standard type of screening material should be used for all pans in a national network.

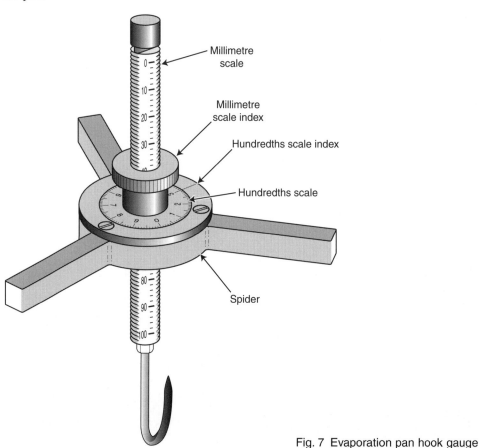

Fig. 7 Evaporation pan hook gauge

78

A screened Class A pan that has been painted white (Sri Lanka).

These points are important because screening reduces the rate of evaporation from a pan by restricting the air flow across the water surface. Data from Sri Lanka (Gunston, 1985) indicated that a screened pan evaporated 33 per cent less than an unscreened pan during a cooler wet season, and 23 per cent less during a hotter dry season. The galvanized wire mesh commonly known as chicken wire mesh is commonly used for screening pans – but be guided by the national practice of the country in which you work.

Pan maintenance to combat the danger of leaks

Apart from drinking birds and animals, evaporation pans will lose water if they leak – once again increasing the apparent evaporation rate. Pans made of galvanized steel sheet are much cheaper to produce than those made of the expensive alloy monel metal. However, when the pan is welded together the galvanized surface is disturbed and a simple equation then applies when the pan is filled with water:

exposed steel surface + water + air = rust

To control rusting, pans made from galvanized steel sheet are usually painted. Rust-preventing primer paints and undercoats are used before the final top coats of paint are applied. Leaks start with very small holes, usually in areas where rusting is obvious, so observers must be thoroughly trained to look out for them.

Pans must be emptied and cleaned at regular intervals. Then is the time to check carefully round the welded seams, and across the floor and wall of the pan, for pin-holes in areas of rust. The observer, or the visiting field team, must be supplied with emery paper to rub down rust spots, together with paints and brushes so that all pan surfaces can be made rust-free and repainted before leaks start to occur. (Do not forget also to supply white spirit – or another cleaning solvent – to clean the brushes after use.)

When an observer or field team find a pan leak of any serious size which has not previously been noticed, field data sheets and station record books must be marked so that allowance can be made in the archived records. Details of pan cleaning and painting should also be recorded in the same way.

National standards for painting pans
Experiments have shown that where pans of the same design installed at the same site are painted in different colours, they can evaporate at different rates. National practices vary. In some countries the outside of pans are painted white, but in other countries aluminium paint is used. The inside may be painted with rust-resisting black bitumen paint. Details will not be discussed here, but – as with pan screening – a national standard for painting pans must be established, and all pan users in the country persuaded to conform to that standard.

Check the pan on each site visit

It has been emphasized before that field teams must check instrument installations thoroughly on each visit to a field station. Evaporation pans should be checked to ensure that:

○ the screening and painting of the pan conforms with the nationally-accepted standards

○ the pan is clean (outside and in)

○ there are no signs of leaks – or of rust spots which could develop into leaks

○ the support timbers and screen (if used) are in good order.

Pans not perfect, but very useful

As mentioned earlier, pans do not provide evaporation estimates which are as accurate as those that can be derived from climatic measurements. However, pan values are very useful for first stage water balance calculations. It is far better to have pan values than no evaporation values at all.

Other types of evaporimeter

Although the Class A type evaporation pan has effectively become an international standard, there are other instruments which estimate evaporation by directly measuring the loss of water stored within them.

Another pan design which has been popular is the Sunken Colorado Pan. This is a square, not a round, pan with sides around one metre long and 500 mm high. The pan is sunk into the ground so that the temperature of the water in it is closer to that of the soil. The scientific arguments between those who favour pans raised above the ground surface (like the Class A), and those who prefer pans sunken in the soil (like the Colorado) will not be discussed here. An important point is the extent to which heat from the sun, stored in the water of the pan, affects the rate of evaporation. However, it is much easier to detect leaks in raised pans than in sunken ones.

Apart from pans, the Piche evaporimeter (sometimes classed as an atmometer) is used to estimate evaporation, especially in countries where French meteorological influence is strong. It consists of a glass tube which is closed at one end and open at the other, very much like a chemical laboratory test tube. A graduated scale is marked along the side of the tube, and a disc of porous paper (like laboratory filter paper) is held over the open end of the tube by a metal clip.

In operation, the tube is filled with purified water and hung, with the filter paper disc at the bottom, inside a met. site thermometer screen (see section on met. instruments, below). As water evaporates from the porous paper disc, the fall in water level is indicated by the graduated scale on the wall of the tube. Like a pan, the Piche evaporimeter is read at a standard time each morning. When the water level falls near to the bottom of the graduated scale on the side of the tube, the Piche is removed from the screen and refilled with water.

The Piche is especially sensitive to clogging of the porous paper disc by surface dust, or chemical salts in the water used. The disc must

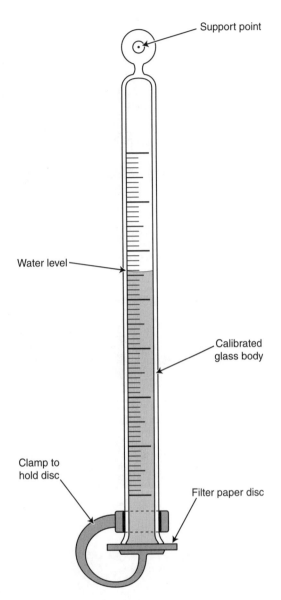

Fig. 8 Piche evaporimeter

be replaced whenever it becomes dirty, which frequently occurs during dusty dry seasons. The water used in a Piche must be as chemically pure as possible. Distilled or deionized water should be obtained from a chemical laboratory if possible.

81

A meteorological station with instruments from which evaporation can be calculated. From left: anemometer (wind run); raingauges; thermometer screen (temperature and humidity); wind vane; sunshine recorder; watchman's shelter (Sri Lanka).

Evaporation equations and climate measurements

The other popular method of obtaining evaporation estimates is to use mathematical equations with input data from climate measurements collected from instruments at met. stations. Because the links between rates of evaporation and climate measurements are complicated, many mathematical equations have been produced to estimate evaporation. However, in the same way that only one design of evaporation pan (the Class A) has been widely accepted internationally, only a limited number of evaporation equations are used extensively world wide.

The FAO Irrigation and Drainage Paper No. 24 *Guidelines for predicting crop water requirements* has already been mentioned as a valuable source of information on practical approaches to estimating evaporation under field conditions. The paper recommends four alternative methods of estimating evaporation from vegetation:

○ evaporation pan
○ the Blaney-Criddle method
○ the radiation method
○ the Penman method.

Of these, evaporation pans have already been discussed and the Blaney-Criddle method estimates evaporation from temperature data. The radiation method is not widely used and the Penman method (which is more accurate) requires measurements of temperature, humidity, sunshine hours (or solar radiation) and wind speed. Since the original research by Dr Penman at the Rothamsted Experimental Station in England, first published in the 1940s, a number of different forms of equations using his basic approach have been developed, but one widely used version is laid out in detail and discussed in FAO 24. The latest, most accurate, version is known as the Penman-Monteith equation.

Estimating evaporation using the Penman method

Although, as just mentioned, a variety of slightly different forms of Penman-type evap-

oration equations have been developed, in most cases the following instruments (installed at a met. station) are needed to collect the required climatic data:

○ air temperature   maximum and minimum thermometers in a thermometer screen
○ humidity   wet- and dry-bulb thermometers in a screen
○ solar radiation   sunshine recorder or electronic recording solarimeter
○ wind speed   cup-counter anemometer.

These measurements relate to a 'manual' met. station, where an observer reads each instrument directly. Automatic weather stations, where climate data are collected and stored electronically, will be discussed later.

Temperature, humidity and the thermometer screen

Thermometers are installed in a ventilated container, known as a thermometer screen, which is mounted on a stand to raise the thermometers to about 1.25 m above the ground. The screen is made of wood, painted white, and is box-shaped. It has double-louvred slatted sides angled and overlapping so that air can pass through, but the instruments within are protected from direct sunlight. The basic installation is of four thermometers, which are read each morning at a time standardized by the national meteorological service (often 09.00 hours). One side of the screen can be opened to gain access to the thermometers.

Maximum thermometer

This is filled with mercury, but has a narrow section close to the bulb (known as a constriction) through which the mercury passes as it expands when the temperature is rising. When the temperature falls, however, the column breaks at the constriction, and the top end of the column in the stem then indicates the highest temperature reached so far.

After the maximum temperature reached

Interior of a thermometer screen. Vertical thermometers are dry bulb (left) and wet bulb with wick (right); horizontal are maximum (upper) and minimum (lower) (England).

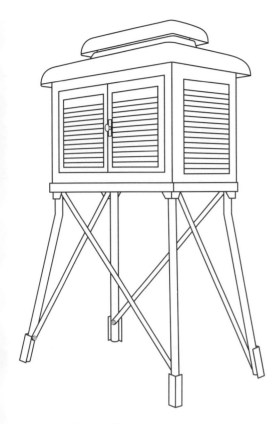

Fig. 9 Thermometer screen

Minimum thermometer

This is not filled with mercury, but with an alcohol-based spirit. A small dumb-bell shaped, dark coloured index is placed within the column of spirit in the glass stem of the thermometer. To set the thermometer, the bulb is tilted up so that the index slides down until it rests against the meniscus at the end of the spirit column in the stem. (The meniscus is the curved surface of a liquid when it is contained in a narrow tube; it is caused by a physical process called surface tension.) The thermometer is then carefully replaced on its mount, which is tilted at about 2° from the horizontal so that the bulb lies slightly below the stem. As the air cools, the spirit contracts and the meniscus moves down the stem towards the bulb, sliding the index with it. When the thermometer is checked each morning, the upper end of the index will lie at the lowest point which the meniscus reached over the past 24 hours, which represents the minimum temperature.

The observer should always read the minimum thermometer by noting the position of the upper end of the index before touching it. The index is very sensitive, and the thermometer need only be tipped gently with its bulb upwards for the index to slide back in contact with the meniscus at the top of the spirit column.

during the past 24 hours has been recorded each morning, the thermometer is reset by swinging it downwards (carefully) with the bulb away from the hand until the mercury column above and below the constriction is continuous. The thermometer is then returned to its mount, which supports it at an angle of about 2° to the horizontal, with the bulb slightly below the stem, to prevent the mercury column in the stem breaking above the constriction. **Warning!** Maximum thermometers need to be swung downwards quite sharply to force the mercury in the stem past the constriction to reset the column. The observer must stand well clear of any obstructions when resetting. It is easy to smash the thermometer on the edge of the screen or on another instrument when shaking it energetically to complete resetting.

Dry and wet bulb thermometers

These are a pair of standard mercury-filled thermometers, which are mounted vertically, side-by-side in the thermometer screen. The bulb at the bottom of the dry bulb is exposed, so that it indicates the air temperature within the screen. The bulb of the wet bulb, however, is totally covered by a cotton wick, the end of which lies within a reservoir of purified water. The continuous evaporation of water from the wetted wick cools the bulb, and the difference in temperature between the dry and wet bulb can be related closely to the humidity of the air within the thermometer screen.

Keep the wet bulb wick clean (and wet!)

During daytime in dry weather, water is evaporating continuously from the wet bulb wick.

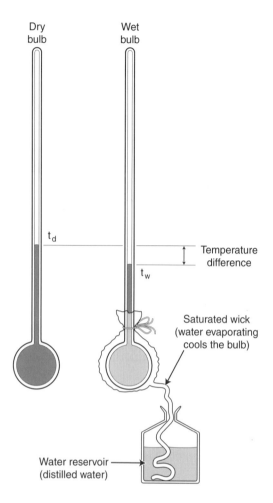

Humidity can be calculated from the difference between dry bulb reading (t_d) and wet bulb reading (t_w).

Fig. 10(i) Dry and wet bulb thermometers

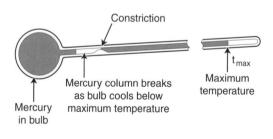

Fig. 10(ii) Maximum thermometer

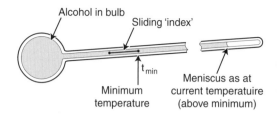

As alcohol in bulb cools, the meniscus at the top of the column slides the 'index' towards the bulb.
As the temperature rises, the 'index' is left in position, indicating the minimum value.

Fig. 10(iii) Minimum thermometer

On hot, dry days the rate of evaporation can be high. Water added to the wick reservoir must therefore be the purest available, so that salts and other deposits dissolved in it are not deposited on the wick after evaporation. If there is a chemical laboratory at headquarters, field teams should arrange to obtain regular supplies of distilled or deionized water to take out to field stations. Water obtained when defrosting refrigerators is also fairly pure. Most important of all, in hot weather the reservoir level must be checked each morning, and extra water added to fill it. If the wick does dry out by mistake, the period for which it was dry must be noted on field data sheets and in the station record book.

Fahrenheit temperature units in old records

It has been the standard practice in most countries over many years for temperature measurements to be recorded in degrees Celsius (°C), with water freezing at 0°C and boiling at 100°C. Older records in archives, however, may be in degrees Fahrenheit (°F), with water freezing at 32°F and boiling at 212°F. If the temperature units are not marked on the archive data (which they should be), the difference in numerical values between °C and °F will usually be large enough for it to be obvious which degree scale applies. Be careful, however, when picking individual temperature values (as opposed to longer sets of readings) from older records.

Solar radiation and sunshine

The amount of energy from the sun which reaches any surface (solar radiation) is a basic driving force causing evaporation. When Penman was developing his evaporation equations in the 1940s, he needed values of solar radiation. However, at that time the only widely used instrument from which such values could be derived was the sunshine recorder, which measured hours of bright sunshine. Although electronic instruments which produce direct values of solar radiation are now more generally available, sunshine recorders are still widely used. In most tropical countries these recorders are installed at major met. stations, and daily sunshine hours values from them are still the most common sources of data used to estimate solar radiation for use with Penman evaporation equations.

Campbell-Stokes sunshine recorder

The principle of operation is simple. A solid glass sphere is used as a lens to focus the sun's rays into an intense spot on a recording card. The card is made of a special paper material which burns only in the immediate area of the focused spot, and is fitted into a curved mount so that the spot moves along the length of the card during daylight hours. When the sun is bright, the heat of the focused spot burns a mark on the card. As long as the sun is shining, a continuous mark is burnt along the card as the sun moves across the sky. If the sun is covered by clouds, however, no mark is burnt on the card.

Each morning the card is carefully replaced by a fresh one. The date of installation must be written on the new card, and the old card must be checked to be sure that the date when it was installed (the previous day) is written on it. Like recorder charts, sunshine cards are useless if they do not carry details of the period for which they were installed. Back in the office, the total length of burnt trace is measured, from which the total number of hours of bright sunshine in the previous 24 hours can be calculated.

The curved metal mount has a series of slots in it, so that at seasons when the sun travels higher or lower in the sky (or when it is north or south of the Equator), cards can be fitted into higher or lower slots, allowing the burnt trace to pass as closely as practical along the centre line of the card.

The sunshine recorder must be installed so that no tall objects (which can cast a shadow) are sited in line between the recorder glass sphere and the sun. A problem with the operation of these recorders is that theft of the glass lens spheres is not uncommon, so extra fixings may be required to hold the recorder down firmly on to its mounting column.

Converting hours of sunshine to solar radiation

Although the sunshine recorder is a simple and reliable instrument, suitable to be installed at met. stations in country-wide networks, the data from the sunshine cards cannot directly produce the values of solar radiation needed for Penman evaporation equations. To establish numerical relationships between sunshine hours and solar radiation, electronic recording solarimeters (discussed below) are installed at met. stations where sunshine recorders are already operated.

Daily totals of sunshine hours and solar radiation are recorded and plotted on a graph. When a sufficient number of daily reading points have been plotted, a 'best fit' straight line is drawn through them. The equation of that line (sometimes called the Angstrom Equation) is then used to convert daily totals of sunshine hours into equivalent daily totals of solar radiation.

This calibration process has not been carried out at every site where a sunshine recorder is installed, but Angstrom Equations have now been prepared for at least one met. station in many countries and climatic regions of the world. Search out and use the version most appropriate for the region of the country in which you are working.

A Campbell-Stokes sunshine recorder. The rays from the sun, focused by the glass sphere, are burning a mark on the sunshine card (Mauritius). (Photograph by David Kirby).

88

Direct recording of solar radiation

Instruments which record solar radiation directly are now more widely used, although they operate on more complicated principles than sunshine recorders. Two of the most popular designs are described below.

Electronic recording solarimeter

This instrument has already been mentioned as providing solar radiation values for direct comparison with sunshine hours data collected at the same met. station site. Solarimeters are also often installed in automatic weather stations. The working centre of the solarimeter is a series of thermocouples, which generate small electrical voltages when exposed to solar radiation. The instrument can produce accurate estimates of radiation, both over short periods, and when summed over 24 hours. However, solarimeters are comparatively expensive in themselves, and the costs of their associated electronic recording systems have to be added. Within tropical countries, therefore, solarimeters tend to be installed only at regional met. centres or at research project sites.

Actinograph

This is a purely mechanical (non-electronic) instrument. It records the temperature difference between a black-coated bimetallic strip exposed to the solar radiation and two similar strips which are either painted white, or shielded from the sun. The bi-metallic strip changes shape as it is heated or cooled. A mechanical linkage transmits the movement to a pen which records the temperature data on a chart that is mounted on a drum and driven by a clock. Daily totals of solar radiation can be derived from the chart traces. Although used for specialist purposes, the actinograph has never achieved country-wide use in national met. station networks.

Solar radiation from daily cloud cover readings

It is normal practice at many met. stations for the observer to record the amount of cloud cover across the sky. This may be done only once a day at the morning reading time, although if afternoon readings are also taken (probably at 15.00 hours), a second cloud cover reading can be recorded then. It is possible to build up an equation linking cloud cover and total daily sunshine hours, and then to use a local Angstrom Equation to estimate daily solar radiation. This is, however, not a very accurate process, so it is not recommended that solar radiation values for use in Penman evaporation equations should be calculated from cloud cover data on a regular basis. However, if a sunshine recorder is out of use for a short period, cloud cover data can be used to estimate missing daily totals of sunshine hours.

Recording wind speed

One of the earliest formal methods of recording wind speed was the Beaufort Wind Scale, which was devised by Admiral Sir Francis Beaufort of the British Navy in 1806. An observer's visual indications of wind speed were given approximate numerical values.

Examples are:

Force 3 (gentle breeze)	Leaves and small twigs in constant motion; wind extends light flag. Equivalent speed at 10 metres above ground within 3.5–5.5 metres per second (8–12 miles per hour).
Force 7 (near gale)	Whole trees in motion; inconvenience felt when walking against the wind. Equivalent speed at 10 metres above ground within 14–17 metres per second (32–38 miles per hour).

The Beaufort scale is very useful in providing estimates of wind speed at the times when regular met. station readings are taken. However, for Penman evaporation estimates values are needed of wind activity throughout 24 hour periods. The most widely used instrument for

A cup-counter anemometer, reading in miles (England).

producing those values is the cup-counter anemometer.

Cup-counter anemometer

This anemometer measures the run of wind, which indicates the movement of air past a fixed point. The anemometer has three cups whose mouths are held in a vertical plane by support arms at 120° intervals round a central vertical collar mounting. The wind drives the cups round, and the vertical collar is connected by a shaft to a mechanical counter/indicator system in the base of the instrument. As the cups turn, the numerical value displayed increases. The shape and positioning of the rotating cups, and the gearing in the counter system, provide an indication of the run of wind in kilometres (or miles) past the anemometer. By reading the anemometer display each morning, the run of wind over the past 24 hours (and therefore the average wind speed) can be calculated, and then used in Penman-type evaporation equations. An alternative design, as used on automatic weather stations, records the rotations electronically.

Kilometres, miles or knots?

As with degrees Celsius or Fahrenheit with temperature readings, always check units when reading anemometers or using archived run of wind data. Older records may well be in miles per hour, although anemometers currently installed at met. stations will normally record wind run in kilometres per hour. Knots (nautical miles per hour) may also be used. One International Nautical Mile measures 1852 metres.

Beaufort scale wind speed data to fill gaps

As with using cloud cover to estimate sunshine hours to fill short gaps in station records, daily (or twice-daily) wind speed estimates obtained using the Beaufort Wind Scale, can be used to fill gaps in wind run records. However, these Beaufort Scale readings will provide much less accurate values than 24 hour totals of wind run from an anemometer.

Look out for growing trees

When discussing rainfall, guidelines were given for ensuring the correct exposure for raingauges. Similar rules apply to siting of anemometers to record wind run. The same comments apply on the distances of high objects from the instrument site and the cutting back of tall vegetation. Dates of cutting back and details of the work done must be recorded in the station record book.

Automatic weather stations

The increasing use of electronic climate instruments was discussed in Chapter 1, where it was noted that installation and maintenance of this type of equipment is carried out by specialist technicians rather than hydrology field teams.

Automatic weather stations are met. stations at which measurements from all the instruments are recorded via electronic signals. Current designs usually store information in solid-state memory modules within data loggers. Earlier designs often stored data on magnetic tape cassettes, like those used to record music. It is also possible for the data to be transmitted through telephone line, radio or satellite telemetry links.

There are a variety of types of automatic weather station, but one originally developed by the Institute of Hydrology in England carried instrument sensors to measure the following climate factors:

○ rainfall (tipping bucket raingauge)
○ dry bulb temperature
○ wet bulb temperature
○ solar radiation
○ net radiation (see below)
○ wind speed
○ wind direction.

Net radiation

This is the balance of incoming solar radiation against radiation emitted from the ground surface. It is measured by a net radiometer, which is basically two solarimeter-type sensors mounted back to back, one pointing up and the other down. Electronic circuits in the recording system add the output signals (incoming solar radiation = +ve; radiation emitted from ground = −ve), producing a net radiation value.

91

An automatic weather station. Instruments are: Kipp solarimeter (on top of pole); wind vane (left) and anemometer (right) on upper arm; net radiometer (left) and temperature screen (right) on lower arm. (The circular tank contains water for the wet bulb thermometer wick.) A tipping bucket raingauge (also connected) stands away from the station framework to the right (Botswana).

Automatic weather stations have been used in a number of hydrological research studies in tropical countries, and they have been installed in remote sites where it is very difficult to operate a conventional 'manual' met. station with a resident observer. The advantages of using automatic stations may be summarized as follows:

○ They are suitable to operate at remote sites where it is difficult to station an observer to read a 'manual' met. station.

○ It is becoming more difficult to find staff who will work as field station observers for relatively low wages in remote areas.

○ Telemetry links (telephone line, radio or satellite) make it easy to collect data, with site visits required only for servicing and maintenance purposes.

○ These instruments can take measurements at very frequent intervals, producing much more accurate data on climate than can be obtained from manual readings by observers, which may be taken only once a day.

○ The prices of some 'manual' station instruments such as sunshine recorders tend to be rising, while the cost of electronic circuit 'chips' – and thus of electronic recording instruments – tends to be falling.

However, as discussed in Chapter 2, the appropriate field hydrology thinking must be applied to the use of automatic weather stations, balancing cost, complexity and reliability against the continuing use of 'manual' met. stations.

Co-operation with national meteorological services

It is useful here to return to a point already discussed in Chapter 1, co-operation between national hydrological and meteorological services. Instruments that measure the climatic factors from which evaporation is estimated are normally installed at a meteorological field station. In many countries, the basic network of met. stations is operated by the national meteorological service, although hydrologists will often use rainfall and climate data from those stations. Reasons why good co-operation between national hydrological and meteorological services (and other organizations which collect hydrological data) is important include:

○ Evaporation pans are frequently operated by organizations other than national met. services.

○ Met. stations may be installed by hydrological and other organizations at sites that are particularly important for water resource management.

○ National met. services often provide training for all met. site observers in a country.

○ Hydrological field teams may be asked to collect data from national network met. stations, and to check that the stations are operating correctly.

Does your organization need to install a met. station?

When hydrologists identify a need for extra climatic data to estimate evaporation in a particular area, they should first check what stations the national met. service or other organizations (such as irrigation schemes or agricultural research centres) already operate in the area. If there is an existing met. station which does not operate all the instruments required to estimate evaporation, it may be possible for the hydrological service to supply and install any extra instrumentation required, but for the met. service observer to read and maintain the new equipment.

However, if your organization does decide to install and operate a new met. station, very useful details can be obtained from the FAO Irrigation and Drainage Paper No. 27 *Agrometeorological field stations*, which was mentioned earlier. The following features are important when laying out the station:

○ To ensure that the best possible readings are taken of rainfall and wind speed, the site should be in a reasonably open situation, with no tall trees, high vegetation or buildings nearby.

○ It should be surrounded by a good fence with a lockable gate.

○ To encourage an even wind flow across the site the fence should have an open structure such as chain link wire mesh.
○ Every attempt should be made to grow grass on the site and to mow it regularly to a short length.

Always visit met. stations in your area

Field programme managers must attempt to visit all the met. stations within the region for which they are responsible, regardless of which organization has the responsibility for operating them. Visits should be arranged in advance to make sure that the station observer will be on site at the time of the visit. Not only will managers discover what instruments are installed at each met. station, but they will make personal contact with the observer. They can also judge the standard of operation and instrument maintenance at the station.

If a met. station lies close to a regular route used by a hydrology field team, the team can call at the station and copy instrument readings directly from the observer's field data sheets and station record book. Arrangements for such visits on a regular basis should be agreed with the station operators and local observers. Regular visits by hydrology teams may also assist the national met. service with their own station inspection programme.

Confused over met. station names?

A confusing variety of names can be used to classify met. stations, for example:

○ meteorological (met.) station
○ climatological station
○ weather station
○ hydrometeorological station
○ agroclimatological station
○ agrometeorological station.

The author prefers to use the overall term meteorological (or met.) station, and then to define the type of station by listing the range of instruments installed there. However, some national met. services use the various station titles in a very precise way to indicate exactly the range of instruments installed. Additional terms like first order and second order stations may also be used.

The subject of met. station classification will not be discussed in detail here, but the prefix 'hydro-' obviously indicates a station which is used to collect data of particular interest to hydrological activities, and 'agro-' likewise for agriculture (including irrigation). 'Met.' and 'weather' in the station title often indicate that data are used immediately to assist national weather forecasting. On the other hand, climatological stations tend to collect data to provide long-term records of weather and climate, suitable for storage in national data archives.

Soil moisture measurement

The amount of water stored within the soil profile, which in general terms is the depth of the soil from which plant roots take up water, can be important in hydrological water balance calculations. The link between transpiration by plants and the availability of water in the soil for roots to take up has already been mentioned. Soil moisture measurement is of particular importance in irrigated agriculture and in the monitoring of research catchments (or watersheds).

One method of calculating the crop factors discussed earlier is to measure the actual losses of water from soil profiles under crops at different stages of growth at individual sites using soil moisture measurement techniques, and then to compare those loss values with rates of evaporation for the same cropped areas, estimated by using evaporation pans or equations.

Hydrological and met. authorities in tropical countries do not normally operate country-wide networks of soil moisture reading stations. Networks of met. stations (producing data from which evaporation can be calculated) provide regional and district values of evaporation. Soil moisture values, although accurate at the point of measurement, can vary greatly over short distances due to variation of soil types and profile depths. Climatic values vary less with distance, particularly where the ground surface is generally flat. Soil moisture meas- urements are therefore better suited where precise information is needed in local areas, and they are widely used to assist irrigation water management.

SECTION 3

SOURCES OF FURTHER INFORMATION

Appendix 1

References and further reading

It has been stressed that this book is an introduction to, not a manual or textbook of, field hydrology. The following list of more detailed textbooks, manuals and reports is a personal one, based on publications that the author has used. It is divided into three broad groups:

○ *International reference works* which aim to cover aspects of field hydrology in tropical countries generally. Some of these come from individual countries, and others from international organizations such as the Food and Agriculture Organization of the United Nations (FAO), the International Organization for Standardization (ISO) and the World Meteorological Organization (WMO). More information on the activities of ISO and WMO appear in Appendix 2.

○ *Country-focused publications* produced by national hydrological and meteorological organizations primarily for use in the country where they have been published. Examples are included from Zambia, East Africa, Sri Lanka, Ecuador, New Zealand, United Kingdom, and the United States. The author's copies of these nationally produced publications are by no means the most recent editions. However, they have been included to indicate the types of field hydrology material which are likely to be available in the reader's own country. Check with senior hydrologists and discover which textbooks and manuals they recommend.

○ *Older textbooks and manuals* which the author has found useful, but which are now generally out of print. They may be available in libraries, or on the bookshelves of senior hydrologists.

International reference works

American Society of Civil Engineers, 1996. *Hydrology handbook* (second edition). Prepared by the Task Committee on Hydrology Handbook. ASCE Manuals and Reports on Engineering Practice No. 28. American Society of Civil Engineers, New York.

Brakensiek, D. L., Osborn, H. B. and Rawls, W. J., 1979. *Field manual for research in agricultural hydrology* (revised February 1979). Agriculture Handbook No. 224, Science and Education Administration (Watershed Hydrology), United States Department of Agriculture, Beltsville, Maryland.

Bos, M. G., Replogle, J. A. and Clemmens, A. J., 1991. *Flow measuring flumes for open channel systems*. American Society of Agricultural Engineers, St. Joseph, Maryland.

Brouwer, C. and Heilbloem, M., 1986. *Irrigation water needs*. FAO Irrigation Water Management Training Manual No. 3, Food and Agriculture Organization of the United Nations, Rome.

Central Board of Irrigation and Power, India., 1987. *Hydrological observations with illustrations* (translated into Hindi and English from an original publication by the Ministry of Construction, Japan). Publication No. 194. Central Board of Irrigation and Power, New Delhi.

Chow, Ven Te, 1964. *Handbook of applied hydrology*. McGraw-Hill Book Co., New York.

Deroo, Georges C., 1981. *Aménagement des parcs agrométéorologiques et règles d'implantation des instruments*. Publications AGRHYMET No. 138.

—— 1984a. *La Pluviométrie; recueil de notes sur les appareils de mesure*. Publications AGRHYMET No. 169.

—— 1984b. *Recueil de notes de cours et d'instructions sur les instruments et les mesures météorologiques.* 2 volumes. Publications AGRHYMET No. 171.

—— 1984c. *La Vapeur d'eau : l'humidité, l'évaporation: recueil de notes sur les appareils de mesure.* Publications AGRHYMET No. 175.

All publications of the Centre Regional de Formation et d'Application en Agrométéorologie et Hydrologie Operationnelle (AGRHYMET), Niamey, Niger.

Doorenbos, J., 1976. *Agro-meteorological field stations.* FAO Irrigation and Drainage Paper No. 27, Food and Agriculture Organization of the United Nations, Rome.

Doorenbos, J. and Pruitt, W. O., 1977. *Guidelines for predicting crop water requirements* (revised 1977). FAO Irrigation and Drainage Paper No. 24, Food and Agriculture Organization of the United Nations, Rome.

Herschy, R.W. (ed.), 1978. *Hydrometry, principles and practices.* John Wiley, Chichester.

Herschy, R.W., 1995. *Streamflow measurement* (second edition). E and F N Spon, London.

Hudson, N.W., 1975. *Field engineering for agricultural development.* Oxford University Press, Oxford.

Hudson, N.W., 1993. *Field measurement of soil erosion and runoff.* FAO Soils Bulletin No. 68, Food and Agriculture Organization of the United Nations, Rome.

International Organization for Standardization. *International standards for liquid flow measurements in open channels.* See Appendix 2.

Jones, G.P. (ed.), 1988. *Lecture notes of the Unesco/Norway fifth regional training course for hydrology technicians, Zimbabwe 1984.* A contribution to the International Hydrology Programme IHP-III, Project 14.1.6. 4 volumes. United Nations Educational, Scientific and Cultural Organization, Paris.

Maidment, D.R. (ed.), 1993. *Handbook of hydrology.* McGraw-Hill, New York.

Meteorological Office, 1982. *Observer's handbook* (fourth edition, second impression). Meteorological Office, Her Majesty's Stationery Office, London.

Miller, S., 1994. *Handbook for agrohydrology.* Natural Resources Institute, Chatham.

Oliver, H.R., 1996. *Hydrological information transfer using HOMS.* UK HOMS National Reference Centre, Wallingford.

Schumacher, E.F., 1974. *Small is beautiful.* Sphere Books, London.

Shaw, E., 1994. *Hydrology in practice* (third edition). E and F N Spon, London.

Smith, M., 1992. *CROPWAT: A computer programme for irrigation planning and management.* FAO Irrigation and Drainage Paper No. 46, Food and Agriculture Organization of the United Nations, Rome.

United Nations Economic Commission for Africa/Transport and Road Research Laboratory, 1990. *The African highway code: A guide for drivers of heavy goods vehicles* (versions available for driving on the left *or* on the right). Transport and Road Research Laboratory, Crowthorne, Berkshire.

United States Bureau of Reclamation, USA, 1997. *Water measurement manual* (third edition). Bureau of Reclamation, United States Department of the Interior, in co-operation with the U.S. Department of Agriculture, United States Government Printing Office, Pittsburgh.

Ward, R. C. and Robinson, M., 1990. *Principles of Hydrology.* McGraw-Hill, London and New York.

World Meteorological Organization, 1994. *Guide to hydrological practices – data acquisition and processing, analysis, forecasting and other applications* (fifth edition). WMO No. 168. World Meteorological Organization, Geneva.

Country-focused publications

Bidwell, L.E., 1971. *Field and office instructions in stream gauging for the Hydrological Survey of Zambia.* United States Geological Survey and the Hydrological Branch, Water Affairs Department, The Zambian Ministry of Rural Development, Lusaka.

Dagg, M., 1968. 'Evaporation pans in East Africa'. in *Proceedings of the Fourth Specialist*

Meeting on Applied Meteorology in East Africa, November 1968, Nairobi.

Edwards, K.A., Gunston, H., Waweru, E.S., 1979. 'The effect of raingauge exposure on catch'. *East African Agricultural and Forestry Journal*, 43, Special Issue – Hydrological Research in East Africa, pp. 289–295.

Fenwick, J. K., 1993 (update of December 1991 edition). *Hydrologist's field manual*. Quality Assurance/Staff Development Unit, Water Resources Survey, DSIR Marine and Freshwater, Christchurch, New Zealand.

Gunston, H., 1985. 'Evaporation pans, and tank and crop evaporation'. in Weller, J.A., Holmes, D.W., and Gunston, H., 1985. *Irrigation management study at Kaudulla, Sri Lanka*. Summary Report, March 1985. H. R. Report No. OD 66. Hydraulics Research, Wallingford.

Gunston, H., 1976. *Report y recomendaciones en la implementation de la red de estaciones 'INERHI' para meteorologia agricola en los distritos de riego mas importantes del Ecuador*. (In Spanish and English.) Instituto Ecuatoriano de Recursos Hidraulicos, the British Ministry of Overseas Development and the Institute of Hydrology, Guayaquil, Ecuador and Wallingford.

Halcrow, Sir William & Partners, 1997. *Site safety in the water industry*. Special publication No. 137. Construction Industry Research and Information Association (CIRIA), London.

Hydrology Division, Sri Lanka, 1974. *The Hydrology branch – its activities and organisational set up*. Hydrology Division, Irrigation Department, Republic of Sri Lanka, Colombo.

Meteorological Office, 1982. *Observer's handbook* (fourth edition). Meteorological Office, Her Majesty's Stationery Office, London.

Pope, R. B., 1992. *Guidance note: Safety in fieldwork*. Natural Environment Research Council, Swindon.

Servicio Nacional de Meteorologia e Hidrologia, Ecuador, 1963. *Instrucciones para la realizacion de la observacion en las estaciones meteorologicas de I and II orden*. Publicacion Serie A, Numero 1. Seccion de Climatologia, Departamento de Meteorologia, Servicio Nacional de Meteorologia e Hidrologia, Ministerio de Fomento, Quito, Ecuador.

Weather Bureau, USA, 1970. *Substation observations: weather bureau observing handbook No. 2*. Data Acquisition Division, Office of Meteorological Operations, Weather Bureau, Environmental Science Services Administration, U. S. Department of Commerce, Silver Spring, Maryland.

Older textbooks and manuals

Ackers, P., White, W. R., Perkins, J. A., Harrison, A. J. M., 1978. *Weirs and flumes for flow measurement*. John Wiley, Chichester.

Andrejanov, V.G., 1975. *Meteorological and hydrological data required in planning the development of water resources (planning and design level)*. Operational Hydrology Report No. 5, WMO No. 419, World Meteorological Organization, Geneva.

Charlton, F.G., 1978. *Measuring flow in open channels: a review of methods*. Report No. 75, Construction Industry Research and Information Association (CIRIA), London.

Dunn, P.D., 1978. *Appropriate Technology – Technology with a human face*. Macmillan, London.

Gangopadhyaya, M., Harbeck, G. E. (Jr.), Nordenson, T. J., Omar, M. H., Uryvaev, V. A. 1966. *Measurement and estimation of evaporation and evapotranspiration*. Technical Note No. 83, WMO No. 201, TP No. 105, Working Group on Evaporation Measurement of the Commission for Instruments and Methods of Observation, World Meteorological Organization, Geneva.

van der Made, J. W., 1986. *Design aspects of hydrological networks*. Proceedings and Information No. 35, Commissie voor Hydrologisch Onderzoek TNO/TNO Committee on Hydrological Research, The Hague.

Winter, E. J., 1974. *Water, soil and the plant*. Science in Horticulture Series, Macmillan, London.

Appendix 2

Field hydrology information from the World Meteorological Organization

The World Meteorological Organization (WMO) is a component body of the United Nations which has the aim of improving standards in meteorology and hydrology worldwide. Most countries of the world are members of the Organization. WMO activities on the collection and analysis of data can be summarized as follows:

> ... to promote standardization of meteorological and hydrological observations and to ensure uniform publication of observations and statistics. With this aim in view, the World Meteorological Congress has adopted Technical Regulations which lay down the meteorological and hydrological practices and procedures to be followed by Member countries of the Organization. These Technical Regulations are supplemented by a number of Guides, which describe in more detail the practices, procedures and specifications which Members are invited to follow ...

Guide to Hydrological Practices
The quotation above comes from the Preface to one of the Guides mentioned, the WMO *Guide to Hydrological Practices*, which has been a major contribution towards improving the practices of field hydrology. The latest (fifth) edition, published in 1994, is a volume of 735 pages, and it is strongly recommended as giving a more detailed coverage of field hydrology

practices than appears in this book. In addition to rainfall, streamflow and evaporation, the Guide covers measurement of snow cover, sediment discharge, soil moisture, groundwater and water quality. Field work practices and safety are covered, as are aspects of hydrological analysis and forecasting. Also included is coverage of the applications of hydrology to water management in such areas as:

○ sustainable water development
○ estimating reservoir capacity
○ flood mitigation
○ irrigation and drainage
○ hydropower
○ navigation and river training.

The Hydrological Operational Multipurpose Subprogramme HOMS has already been mentioned in Chapter 1, and may be summarized as follows:

> (HOMS) ... consists of an international network for the transfer of packages of proven hydrological know-how, with particular reference to helping developing countries. More than one hundred HOMS Centres worldwide produce and supply information components, and handle requests for components for use in their own countries.

The quotation, together with other material on which this chapter is based, comes from a report titled *Hydrological information transfer using HOMS*, prepared by Dr Howard R Oliver, National Representative for HOMS in the United Kingdom.

Member countries of WMO who wish to participate in HOMS do so by setting up a HOMS national reference centre. These centres are usually operated by national hydrological services, and each centre is led by a named HOMS national representative, who can be contacted to provide full details of the availability of HOMS components.

HOMS components

These are designed as self-contained units of technical information, so that each can 'stand alone' as a source of advice on a particular topic. Different components may be manuals, computer programs, instructional video tapes or other forms of information storage. Summary information on each component is included in the *HOMS Reference Manual*, a copy of which is held at each national reference centre. Components can be of different levels of complexity, and they may be grouped in HOMS sequences – where a number of individual components are used together.

Classification of HOMS components

The basic classification under which HOMS components are grouped is as follows:

o policy, planning and organization
o network design
o instruments and equipment
o remote sensing
o methods of observation
o data transmission
o data storage, retrieval and dissemination
o primary data processing
o secondary data processing
o hydrological forecasting models
o hydrological analysis for the planning and design of engineering structures and water-resources systems
o groundwater
o mathematics and statistical computations
o training aids in operational hydrology.

The components that have been contributed to build up HOMS come from many member countries. Each component has to meet the standards and recommendations laid down in the *Guide to Hydrological Practices* and also in *WMO Technical Regulations*. A very useful

feature of the latest edition of the *Guide to Hydrological Practices* is that topics are cross-referenced to the *HOMS Reference Manual*.

Details of the activities of WMO, together with a catalogue listing all the publications of WMO that are currently available, can be obtained from:

The Secretary-General
World Meteorological Organization
P.O.Box 2300
CH-1211 GENEVA 2
Switzerland

Details can also be obtained from WMO offices in individual member countries.

The *Standards for Liquid Flow Measurements in Open Channels* which have been prepared through the International Organization for Standardization (ISO) have already been mentioned. They are very important documents as they give full and precise descriptions and instructions relating to flow measurement techniques, based on international agreement between hydrometric experts from many countries of the world. As discussed earlier, by using the internationally-agreed methods of working and construction detailed in these Standards, the same hydrometric flow measurement activity can be carried out in the same way anywhere in the world.

A current list of those Standards follows. 'ISO' usually indicates that the publication is an ISO International Standard. 'ISO/TR' denotes an ISO Technical Report; these are either interim progress reports or they contain factual information of a kind different from that normally incorporated in an International Standard.

Where the full phrase 'Liquid flow measurement in open channels' has been used in the formal title of a Standard, it has been abbreviated here to '*Lfmioc*'; likewise '*Mlfioc*' has been used instead of the full title phrase 'Measurement of liquid flow in open channels'. However, when ordering ISO Standards, these titles should be used in full.

The International Organization for Standardization

Copies of these Standards can be obtained through the national office of the ISO in the country where you are working. Further information can be obtained from:

Sales Department
ISO Central Secretariat
1, rue de Varembé
Case postale 56
CH-1211 Geneva
Switzerland

17.120.20 Flow in open channels

ISO 748:1997
Measurement of liquid flow in open channels (*Mlfioc*) – Velocity–area methods

ISO 772:1996
Hydrometric determinations – Vocabulary and symbols

ISO 1070:1992
Liquid flow measurement in open channels (*Lfmioc*) – Slope–area method (plus AMD 1 – Amendment 1:1997)

ISO 1088:1985
Lfmioc – Velocity–area methods – Collection and processing of data for determination of errors in measurement

ISO 1100-1:1996
Lfmioc – Part 1: Establishment and operation of a gauging station

ISO 1100-2:1982
Lfmioc – Part 2: Determination of the stage–discharge relation

ISO 1438-1:1980
Water flow measurement in open channels using weirs and Venturi flumes – Part 1: Thin-plate weirs

ISO 2425:1974
Measurement of flow in tidal channels (plus AMD 1 – Amendment 1:1982)

ISO 2537:1988
Lfmioc – Rotating element current-meters

ISO 3454:1983
Lfmioc – Direct depth sounding and suspension equipment

ISO 3455:1976
Lfmioc – Calibration of rotating-element current-meters in straight open tanks

ISO 3716:1977
Lfmioc – Functional requirements and characteristics of suspended sediment load samplers

ISO 3846:1989
Lfmioc by weirs and flumes – Rectangular broad-crested weirs

ISO 3847:1977
Lfmioc by weirs and flumes – End-depth method for estimation of flow in rectangular channels with a free overfall

ISO 4359:1983
Lfmioc – Rectangular, trapezoidal and U-shaped flumes

ISO 4360:1984
Lfmioc by weirs and flumes – Triangular profile weirs

ISO 4362:1992
Mlfioc – Trapezoidal profile weirs.

ISO 4363:1993
Mlfioc – Methods for measurement of suspended sediment

ISO 4364:1997
Mlfioc – Bed material sampling

ISO 4365:1985
Liquid flow in open channels – Sediment in streams and canals – Determination of concentration, particle size distribution and relative density

ISO 4369:1979
Mlfioc – Moving-boat method

ISO 4371:1984
Mlfioc by weirs and flumes – End-depth method for estimation of flow in non-rectangular channels with a free overfall (approximate method)

ISO 4373:1995
Mlfioc – Water level measuring devices

ISO 4374:1990
Lfmioc – Round-nose horizontal broad-crested weirs

ISO 4375:1979
Mlfioc– Cableway system for stream gauging

ISO 4377:1990
Lfmioc – Flat-V weirs

ISO 6416:1992
Mlfioc – Measurement of discharge by the ultrasonic (acoustic) method

ISO 6420: 1984
Lfmioc – Position fixing equipment for hydrometric boats

ISO/TR 7178:1983
Lfmioc – Velocity–area methods – Investigation of total error

ISO 8333:1985
Lfmioc by weirs and flumes – V-shaped broad-crested weirs

ISO 8363:1997
Mlfioc – General guidelines for the selection of methods

ISO 8368:1985
Lfmioc – Guidelines for the selection of flow gauging structures

ISO/TR 9123:1986
Lfmioc – Stage–fall–discharge relations

ISO 9195:1992
Lfmioc – Sampling and analysis of gravel-bed material

ISO 9196:1992
L*fmioc* – Flow measurements under ice conditions

ISO/TR 9209:1989
Mlfioc – Determination of the wetline correction

ISO/TR 9210:1992
Mlfioc – Measurment in meandering rivers and in streams unstable boundaries

ISO/TR 9212:1992
Mlfioc – Methods of measurement of bedload discharge

ISO 9213:1992
Measurement of total discharge in open channels – Electromagnetic method using a full-channel-width coil

ISO 9555-1:1994
Mlfioc – Tracer dilution methods for the measurement of steady flow – Part 1: General

ISO 9555-2:1992
Mlfioc – Tracer dilution methods for the measurement of steady flow – Part 2: Radioactive tracers

ISO 9555-3:1992
Mlfioc – Tracer dilution methods for the measurement of steady flow – Part 3: Chemical tracers

ISO 9555-4:1992
Mlfioc – Tracer dilution methods for the measurement of steady flow – Part 4: Fluorescent tracers

ISO/TR 9823:1990
Lfmioc – Velocity–area method using a restricted number of verticals

ISO 9825:1994
Mlfioc – Field measurement of discharge in large rivers and floods

ISO 9826:1992
Mlfioc – Parshall and SANIIRI flumes

ISO 9827:1994
Mlfioc – Streamlined triangular profile weirs

ISO/TR 11328:1994
Mlfioc – Equipment for the measurement of discharge under ice conditions

ISO/TR 11655:1995
Mlfioc – Method of specifying performance of hydrometric equipment

ISO/TR 11656:1993
Mlfioc – Mixing length of a tracer

ISO/TR 11974:1997
Mlfioc – Electromagnetic current meters

Appendix 3

Measurement units and equations

Thin plate weirs

1. V-notch weir The most commonly used type has a 90–degree notch, with each side at 45 degrees to the vertical. The following table, from *Field engineering for agricultural development* by N. W. Hudson (see Appendix 1) relates discharge to stage or head for a 90-degree V-notch weir:

Head/stage (mm)	Discharge ($l.s^{-1}$)	Head/stage (mm)	Discharge ($l.s^{-1}$)
40	0.44	180	18.9
50	0.73	190	21.7
60	1.21	200	24.7
70	1.79	210	27.9
80	2.49	220	31.3
90	3.34	230	35.1
100	4.36	240	38.9
110	5.54	250	43.1
120	6.91	260	47.6
130	8.41	270	52.3
140	10.2	280	57.3
150	12.0	290	62.5
160	14.1	300	68.0
170	16.4	350	100.0

Stage is only quoted to 300 mm as V-notch thin plate weirs are commonly used to measure comparatively small discharges.

'Half-' and 'quarter-90 degree' V-notch weirs Weirs with angles less than 90 degrees can be used, two types being described as 'half-' and 'quarter-90 degree'. However, these classifications are confusing, as the 'half' and 'quarter' refer to reduction of discharge (as compared with a 90 degree weir) for the same head or stage, and not to the measured angle of the V-notch. Using the table above, the dischar-ges for these notches at a given stage would be half or quarter of the values listed.

The angle of a 'half 90 degree' notch is 53 degrees 8 minutes (not 45 degrees) and, for a 'quarter 90 degree' notch, the angle is 28 degrees 4 minutes (not 22 degrees 30 minutes).

2. Rectangular weirs Providing a rectangular thin plate weir is *fully contracted* (see description in Chapter 5), the following table (also from Hudson) relates head or stage to flow per metre of crest length:

Head/stage (mm)	Discharge per metre of crest length ($l.s^{-1}$)	Head/stage (mm)	Discharge per metre of crest length ($l.s^{-1}$)
30	9.5	210	169.5
40	14.6	220	181.5
50	20.4	230	193.5
60	26.7	240	205.5
70	33.6	250	218.5
80	40.9	260	231.0
90	48.9	270	244.0
100	57.0	280	257.5
110	65.6	290	271.0
120	74.7	300	284.0
130	84.0	310	298.0
140	93.7	320	311.5
150	103.8	330	326.0
160	114.0	340	340.0
170	124.5	350	354.0
180	136.0	360	368.5
190	146.0	370	383.5
200	158.5	380	398.0

The slope–area method of estimating streamflow and Manning's equation

The slope–area method of estimating discharge in ungauged channels is discussed in Chapter 5; it is especially useful for estimating flood flows. The method is normally based on the Manning equation for flow in open channels, which relates velocity and discharge to measurements linked to the channel cross-section and slope. A *roughness coefficient* ('*n*') – commonly known as 'Manning's "n"'– is used, which relates to the resistance offered to flow by the surfaces of the channel bed (e.g. large rocks or extensive weed growth).

The Manning equation

Velocity $\quad V = n^{-1}.R^{2/3}.S^{1/2}$
Discharge $\quad Q = n^{-1}.A.R^{2/3}.S^{1/2}$
where: $\quad$ V = mean flow velocity (m.s^{-1})
$\qquad\quad$ Q = discharge (m^3.s^{-1})
$\qquad\quad$ n = roughness coefficient
$\qquad\qquad$ (Manning's '*n*')
$\qquad\quad$ R = hydraulic radius = A/P (m)
$\qquad\quad$ A = mean cross-section area (m^2)
$\qquad\quad$ P = wetted perimeter (m)
$\qquad\quad$ S = slope of water surface
$\qquad\qquad$ (m.m^{-1})

The wetted perimeter is the distance along the bed of the channel (at right angles to the direction of flow) between the points where the water surface touches either bank along the line where the channel cross-section is estimated.

The method of operation is to select two or more cross–sections along a chosen river reach measuring site in advance of likely floods, and to mark their positions on the banks (above likely flood levels) with stakes. The cross–secton and wetted perimeter are measured during periods of flow. At the time of flow measurement itself, a surveying staff and level are required to indicate the slope of the water surface between the selected cross–sections.

According to Fenwick, in the New Zealand *Hydrologist's Field Manual*, a river reach selected for use with the slope–area method should:

○ be straight
○ contain the flow without overflow at the stages measured
○ have a flood waterway obstructed by a minimum of vegetation
○ have uniform cross-sections and preferable be converging (i.e. with cross-sections becoming progressively smaller in area downstream); the increasing velocity through the reach prevents deposition of bed load and ensures a stable bed profile
○ be sufficiently long that uncertainties in slope measurements will not be significant
○ have floods marks of good quality and quantity, in the case of measurements (of flood stage at cross-sections and of channel slope) made following an event
○ be sufficiently far away from sharp bends, either above or below, so that water levels on each bank will not differ greatly in height, and significantly backwaters do not occur.

For natural stream channels, Hudson quotes the following values for Manning's 'n':

Clean and straight	0.025–0.030
Winding, with pools and shoals	0.033–0.040
Very weedy, winding and overgrown	0.075–0.150

For streamflow estimation it is best to use 'clean and straight' river reaches. For smooth channel beds a value of 0.020 may be used for 'n' but 0.030 should be used for rough, rocky beds on straight reaches. Fenwick notes that for gravel bed rivers 'it is typically in the range 0.020 to 0.050, and tends to vary with discharge.'

Only an outline of the slope–area method has been given. Considerable practical skill is required to use it well to obtain reasonable estimates of streamflow. Advice on its use should be requested from senior hydrologists.

Conversion factors

The Conversion factors between Imperial (British) to metric units covering water quantity and discharge, as mentioned in the section

Flow measurement unit confusion: cumecs, cusecs and mgd in Chapter 5 (page 68), are as follows:

1 cumec = 35.3 cusecs (or second-feet)
1 mgd (Imp) = 4.55 tcmd (or Ml/d)
1 acre-foot = 1233 cubic metres (m^3)

A further reminder of the difference between British Imperial and US gallons:

1 British 'Imperial' gallon = 4.55 litres
1 US gallon = 3.78 litres
1 US gallon = 0.83 Imperial gallons
1 Imperial gallon = 1.20 US gallons